Cooperatives and Development
Agricultural Politics in Ghana and Uganda

Cooperatives

&

Development

Agricultural Politics
in
Ghana and Uganda

Crawford Young
Neal P. Sherman
Tim H. Rose

The University of Wisconsin Press

Published 1981

The University of Wisconsin Press
114 North Murray Street
Madison, Wisconsin 53715

The University of Wisconsin Press, Ltd.
1 Gower Street
London WC1E 6HA, England

First printing

Printed in the United States of America

For LC CIP information see page 276

ISBN 0-299-08710-7

Preface

The stately Ugandan elephant must step aside for this manuscript. Nothing in the animal kingdom compares to its extraordinary gestation period: no less than 15 years. In 1965, when one
of the authors (Young) was invited to spend a year at Makerere
University (Uganda) as visiting professor, this study first took
shape.

Initially, the study was to focus upon the cotton and coffee
cooperatives in Uganda alone. It was undertaken in collaboration
with Professor E.A. Brett, then on the academic staff of Makerere
University, now of Sussex University (United Kingdom). During
the 1965-66 academic year, we began the interviews, document collection, and the survey of some 500 Ugandan farmers described in
Chapter 5. Brett then continued the field inquiry in 1966-67,
finishing up on the administration of the survey.

For some years, the study then lay fallow. Brett completed
other work on the political economy of East Africa (most notably
his fine study, Colonialism and Underdevelopment in East Africa
[New York: Nok Publishers, 1973]). Young went through a period
of administrative service at the University of Wisconsin-Madison.
Distance and other factors made completion of the study on the
original collaborative basis impossible. Indeed, for a time it
appeared that the considerable body of material amassed would
simply be plowed under and forgotten.

However, the possibility for recurrection of the project on
a redesigned basis emerged in the early 1970s, when Neal Sherman
and Tim Rose, then doctoral candidates at the University of Wisconsin-Madison, chose dissertation topics which bore upon the
theme of agricultural cooperation. Sherman carried out research

in Uganda during the 1971-73 period, concerning the government role in the emergence of a significant dairy industry. In the course of his inquiry, he collected extensive materials relating to the dairy cooperatives. Rose was in Ghana in 1973-75, studying the cocoa cooperatives. Both Sherman and Rose, as a part of their research, undertook surveys of dairy and cocoa farmers, respectively, using a modified version of the questionnaire employed by Brett and Young.

Young, Sherman, and Rose then decided to pool their efforts in the present study, which thus became a comparative inquiry into the political economy of rural cooperation in Uganda and Ghana. While the new concept of the manuscript took form in 1975, it was not until 1978-79 that the actual drafting occurred. By that time, Young was still at the University of Wisconsin-Madison, but Sherman taught at Tel-Aviv University in Israel, and Rose was employed by Morgan-Newman Associates in Washington, a consulting firm specializing in overseas development. The composition of the manuscript thus occurred while we were at a great distance from one another. However, we did have two occasions for collective consultation: in New York, in September 1978, when most of the first draft was complete, and initial critical comments were available; and in Madison, in April 1980, when we had available several detailed critiques of the initial draft. Chapters 1, 3, 4, 5, 6, and 9 were initially drafted by Young; 2, 7, 8, and 13 by Sherman; and 10, 11, and 12 by Rose. All of us then contributed to the process of editorial revision.

For a project extending over so many years, an unabridged listing of the innumerable persons, groups, and organizations who offered their help and support would be of unseemly length. While our heartfelt thanks, as well as the customary exemption from any share of the blame for whatever deficiencies may be noted in this study, go to every single one who played any part, however small, in making the book possible, we can mention only a fraction.

We begin with our material benefactors. At different stages in this study, one or another of us received financial support from the Rockefeller Foundation, the Social Science Research Council, the Foreign Area Fellowship Program, the Midwest Universities Consortium for International Activities (MUCIA), the University of Wisconsin Graduate School Research Committee, and the Center for Development at the University of Wisconsin. Final editorial work was completed by Young while a Visiting Member of the Institute for Advanced Study at Princeton.

Publication costs for the book were met in part by grant funds available to the University of Wisconsin Press. The University of Wisconsin Land Tenure Center helped make publication possible, by its sponsorship of the book. Their editorial and technical services were of inestimable help in preparing the manuscript for printing. Maps were prepared by the University of Wisconsin Cartographic Laboratory.

     Brett, as initial collaborator on the Uganda part of this
venture, contributed in many ways to its early conceptualization,
as well as to the data collection. Our intellectual debt to him
is particularly large. Two specialists in agricultural economy,
Deryke Belshaw and Malcolm Hall, were unflagging in their efforts
to raise our level of understanding of the technical and economic
aspects of agriculture. Ian Wallace and Dennis Skene generously
made available to us their own field research materials from Bugisu and Bunyoro, respectively. James Coleman and William H.
Young provided timely support and encouragement at particular
crucial junctures.
     We are grateful for the assistance of Edward Takhuli, Samuel
Busulwa, James Katorobo, and Isaac Ojok in the administration of
the Uganda survey. Marshall Carter, Kathleen Lockard, Catharine
Newbury, and Thomas Turner helped us code and analyze the survey
results.
     A number of colleagues did their utmost to make this a bet-
ter manuscript than it was initially. Special recognition is ow-
ing Nelson Kasfir, for his extraordinarily detailed, perceptive,
and comprehensive critique, which made clear to us a number of
important shortcomings. Others who offered valuable criticisms
include Michael Schatzberg, Fred Hayward, Dennis Dresang, James
Pletcher, Edward Bewayo, Okello Oculi, Victor Le Vine, and John
Bennett.
     We were fortunate to enjoy the full cooperation of the Coop-
erative Departments in both Uganda and Ghana for our study. In
addition, a number of cooperative organizations in Uganda were
gracious with their assistance; so also was the Ghana Cooperative
Marketing Association, and Ghana Cooperative Alliance, and the
Produce Buying Agency in Ghana. Makerere University, Kampala,
and the University of Ghana made both their intellectual and
their physical resources available to us.
     The final step in the 15-year pathway from research dream
to published book is the index; for this, we owe our thanks to
Eva Young for her patient and dedicated work.

                         Crawford Young, Neal Sherman, Tim Rose
                         November 1980

Contents

List of Maps

Cooperatives and Development
Agricultural Politics in Ghana and Uganda

1. Cooperatives in Rural Development: Comparative Lessons of Experience

It all began in a Swiss cheese factory.

Cooperative genealogy, for those who wish to give it an antique pedigree, can be traced in germ back to thirteenth-century Swiss cheese makers, banding together in the production and marketing process.[1] However, as a significant institution, cooperatives date only from the nineteenth century, and the multiple transformations of society united to the industrial revolution: the rise of capitalism, and reactions to it, in particular various strands of socialist thought; the spread of commercial agriculture, and technological changes in rural production. As a modern phenomenon, cooperatives originated in England, primarily as urban, consumer retail enterprises. Later in the nineteenth century, the cooperative idea spread to the rural sector, acquiring real importance in northern Europe and the plains and prairies of North America. With the triumph of socialist revolutions in Russia and China, and the forced spread of socialism in Eastern Europe, the cooperative--quite differently conceived--became an instrument of the imposition of socialist agriculture on the peasantry. After World War II, the new state of Israel, with a strong core of settler-farmers of high ideological motivation, organized its agrarian economy on a predominantly cooperative basis. In Latin America, cooperatives were one of the formulas pursued by reformers intent on breaking the latifundia tradition. Finally, many of the newly independent nations of Africa and Asia saw in cooperatives a multipurpose vehicle for achieving a broad array of state objectives. For a number of these, the attraction of cooperation lay partly in its compatibility with a broadly (if often diffusely) anti-capitalist perspective.

-3-

This capsulized résumé makes clear that the umbrella term "cooperation" covers a wide variety of particular forms. In its origins, cooperation was an anti-capitalist social movement; in its recent applications, it has more frequently served as an instrument of the state for structuring the rural sector. The normative doctrine of cooperation remains heavily influenced by the perspectives of the nineteenth-century founders of the movement. Contemporary practice is likely, in reality, to widely diverge from the formal ethos. This contradiction may help us understand a number of seeming paradoxes: the voluntaristic, participatory ethic versus the practical reality of routinization and bureaucratic oligarchies in the older cooperatives of Northern Europe and North America;[2] pervasive state regulation in the newer third world cooperatives; the frequently expressed enthusiasm for the formula, contrasted to the many observed shortcomings of the movement in practice.

FOCUS OF THE STUDY

In this study, we propose a comparative inquiry into the cooperative experience in two African states, Uganda and Ghana. We explore the role, accomplishments, and setbacks of rural cooperatives, both as a policy instrument for the promotion of agricultural expansion, and as a vehicle for achieving goals desired by their farmer participants. In both cases, the only significant form of cooperative organization was the agricultural marketing society. The great majority of cooperative membership was linked to the marketing of export crops: cotton and coffee in Uganda, and cocoa in Ghana. In both countries, cooperative participation--at its peak--was very large: over 500,000 in Uganda, and 200,000 in Ghana.

The comparative frame rests easily upon these polities, because they are strikingly similar in a number of respects. The two countries are remarkably identical in scale--92,000 square miles for Ghana, and 94,000 for Uganda; 8,546,000 population in 1970 for Ghana, and 9,760,000 the same year for Uganda. Both achieved identity as states under British rule; the broad lines of colonial policy were quite parallel. From an early date, both territories became heavily dependent upon one or two export crops from which the state derived the bulk of its revenue. A racial division of labor obtained in these crops, with African farmer producers, and expatriate processors and exporters. The sequences of post-independence politics were similar as well. In an initial period, both governments carried forward the liberal planned economy policies of the late colonial era. Both briefly experimented with a more socialist orientation (1961-66 in Ghana, 1969-70 in Uganda), which was cut short by military intervention. Then, in the 1970s, the apparatus of the state, and its capacity to foster development of any ideological inspiration, was sharply corroded--in Ghana, by the extraordinary and predatory corruption of the Acheampong regime, and in Uganda by the capricious tyranny of the Amin years.

While the comparabilities of context stand out, there were also important differences in the life histories of cooperation in the two countries. Though in both cases cooperative development received some paternal nurture by the withdrawing colonial administration, its relationships with the nationalist political leadership differed sharply. In Uganda, the cooperative movement was offered strong official backing, and eventually awarded by fiat a marketing and processing monopoly in cotton and coffee. In Ghana, the hero of the independence movement and first leader, Kwame Nkrumah, always viewed the cooperatives with suspicion, perhaps partly because their centers of strength were in the most dissident region of the country. In 1961, they were dissolved, and their assets confiscated in favor of an organ of his ruling party, the Convention Peoples Party (CPP). The remnants of the cooperatives then rose from their ashes after the fall of the Nkrumah regime, with benign tolerance, but less than total support from the military successors. After a brief resurgence, a new decline set in, this time mainly from internal shortcomings, and the cooperatives were again suppressed in 1977. In Uganda, the formal shell of cooperation survived even the devastation of the Amin years.

While the life histories of the agricultural cooperatives contrasted in the two countries, they were in both states of sufficient importance in the rural political economy to offer a useful insight into the institutional interfaces between the smallholder and the polity. In most African states, the rural sector has proved the Achilles' heel of development programs. As Guy Hunter has observed, whatever the elegance of the five-year plans produced in the capital, higher production can only come from millions of smallholders, who may be submissive and quiescent politically, but who remain autonomous in their economic decisions on their plots.[3] Whether the explanation lies in the failure of either the colonial or post-colonial state to "capture the peasantry" and eliminate the "peasant mode of production," as argued by Hyden,[4] or the perverse structuring of market incentives which drives smallholders from the crops the state promotes, as Bates suggests,[5] the outcome is the same: agricultural stagnation. An examination of the particular experience of the rural cooperatives will help illuminate the unresolved dilemmas in the relationship between the small producer and the broader national system.

To locate our study historically and comparatively, we will begin with a brief review of the cooperative experience, to trace the evolution of its institutional ideology. We will then explore at a theoretical level the promise and contradictions of

the cooperative formula in third world rural settings, drawing
upon the wide range of comparative literature now available. We
turn then to consider in detail the Ugandan and Ghanaian cases.
In Uganda, we examine the cotton and coffee cooperatives, and
also the recent venture in dairy cooperation. In Ghana, we
trace the rise and fall of the cocoa cooperatives. Finally, in
a concluding section, we return to a systematic comparison of
these two experiences.

PRINCIPLES OF COOPERATION:
EVOLUTION OF THE INSTITUTIONAL IDEOLOGY

At an official level, the world cooperative movement, represented by the International Cooperative Alliance, has defined
the irreducible set of principles which define cooperation as
open membership, democratic control by members, relative equality of share capital, and return to members proportionate to
business done through the cooperative.[6] The ICA, in reality,
tends to reflect the perspectives of the older segments of the
cooperative movement. They reflect very imperfectly, however,
the operating premises of organizations bearing the cooperative
designation in the third world whose spokesmen declare that a
paternal state of tutelage is indispensable for the nurture and
growth of cooperatives in Africa and Asia.[7] To situate some
of the tensions between ideology and practice in contemporary
cooperatives, a backward glance at the evolution of this organizational device is appropriate.

Precursors of the cooperative ideology may be found in the
radical ferment of seventeenth-century England. John Ballers, a
Quaker, wrote a tract in 1695 entitled "Proposals for Raising a
colledge of Industry for all Usefull Trades and Husbandry," which
was a blueprint for building a direct relationship between producers and consumers on a democratic and egalitarian basis, eliminating the "middleman."[8] However, it was the full-blooded
emergence of the capitalist era which generated the idea of cooperation as an alternative economic ethic.

The high idealism of the Owenite socialists offered a framework for the cooperative movement. Owenite philosophy was suf-
fused with the Rousseauvian premise that man is innately good,
and has been corrupted only by the institutions which enclose
his existence. The heavenly mill town of New Lanark in Scotland
in the early nineteenth century, managed with celestial paternalism by Robert Owens, spawned a whole wave of experimental communities, especially in North America, where the effort was made
to let the natural goodness of man find full reflection in egalitarian and cooperative institutions.

While enterprises seeking to base the entire social existence of a small collectivity on communitarian principles almost
invariably failed,[9] the Owenite principles fared much better

in limited application in the economic realm. In 1844, the Rochdale Cooperative Society was formed, its founders immortalized in cooperative hagiography as the Rochdale pioneers. They set the model for the consumer cooperative. A store was established, with members pooling their labor and working capital. The surplus earned was then returned in the form of dividends to the members. Strict rules of equality were maintained; members each had one vote in the management of the store, and limits were set to their financial participation. The cooperative was, in short, an economic enterprise founded upon the ideals of mutual help and responsibility, democracy, equality, and freedom. Although the functions it performed were no different from those of another shop, the ethic which governed its operations was; the banal pursuit of commerce was, for the founding generation, ennobled by the ethos of cooperation which inspired its operation. Success required the happy by-product of more inexpensive acquisition of the goods handled by the consumer cooperative; the ethic could not be expected to withstand actual economic loss. But a reasonably well-managed cooperative was a living morality play; in a small way, the cooperators rewarded virtue, and by their very existence acted out a living reproach to the greed, avarice, and exploitation which were held to inhere in the economic institutions of industrial capitalism.

The cooperative movement in Britain carved out a modest but respectable place, primarily through the expansion of consumer cooperatives. By the postwar era, cooperatives controlled 30,000 shops, and numbered 12 million members; 75 percent of their trade was in foodstuffs.[10] In recent years, however, the consumer cooperative has been a declining force, unable to compare in service or price with the supermarket, and with its sense of social purpose eroded more by the growing dominance of its managers. Indeed, one recent study of British cooperatives concluded that the divorce between ownership (members) and control (managers) was greater today than in corporate enterprises.[11]

In northern Europe--Scandinavia and Germany--the cooperative idea was first translated to the agricultural sector on a large scale. For the marketing of major crops--grains, meat, dairy produce--and selling farming inputs, such as fertilizer and implements, cooperatives provided an alternative to the commercial intermediaries tied to the hostile urban world. In Germany, in the late nineteenth century, Friedrich Raiffeisen led the cooperatives in a major new direction through the administration of rural credit to finance farm operations. In Ireland cooperatives were at first a quite potent social vehicle for Irish farm families against English mercantile and landlord interests. In Denmark, cooperatives served farmers in their struggle against the old landlord aristocracy. Sweden now has an enormous national cooperative, with 1 million members, and annual turnover of over $500,000,000.

In North America, cooperatives flourished primarily in the

farm sector; regionally, they were concentrated in the Middle West, and prairie states and provinces of the United States and Canada. Here Scandinavians and Germans formed an important component of the rural population; to many, the cooperative formula was already familiar.   Here also the populist impulses ran strongly; the eastern bankers and railway magnates, through their control of the marketing mechanisms, had a powerful commercial advantage which they ruthlessly exploited.   Dairy cooperatives in the United States have become a billion dollar undertaking; the Chicago Producers' Cooperative Association is the largest single dealer on the Chicago livestock market.   Grain elevators are largely cooperatively owned; access to adequate storage is critical to protecting the market position of the farmer at harvest time.   Perhaps the most eloquent commentary on the changing nature of cooperatives in the American scene was the threatened anti-trust prosecution of the largest dairy cooperative, The Associated Milk Producers, in 1972.   A timely contribution to the Nixon re-election fund ended the prosecution threat, but not the irony of a cooperative whose scope and style of operation qualified it for trust-busting.[12]

A life cycle is clearly discernible in the well-established cooperatives of the industrial world.   In the beginning, a burst of moral energy was captured by the new institutions.   Cooperation was a solidary riposte to the predatory forms of the capitalist economy.   The movement freed the farmer from the sense of helpless dependency on urban commercial interests for marketing, storage, or supply of key inputs.   Participation in the early phase is high: the mundane execution of economic tasks is invested with social purpose.   However, once successfully launched, the very effectiveness of the cooperatives in filling an economic niche creates a new set of imperatives.   To survive, the cooperative must become efficient.   For a mere social or political organization, efficiency has no real empirical measure; for the cooperative, the balance sheet offers its constraining reckoning on a regular basis.   While cooperation is an ethos, efficiency is the incubator of technocracy.   Yet cooperatives could not in the long run survive in a liberal economy, if the services they rendered their members were available more cheaply on the open market.

The implications of this simple fact are many.   As cooperatives achieved a certain scale, they could no longer be directly managed by their members, but had to hire specialized managerial staff.   Armed with the efficiency criterion, the managerial cadres tended to enlarge their role, while the representative organs of the cooperative tended to atrophy; the "iron law of oligarchy" detected by Roberto Michels in labor unions and socialist parties had its analogues in the cooperatives.   As cooperatives become institutionalized, they become primarily economic agencies, operated by specialized managers under the discipline of the market, with effective member participation only a residual phenomenon, and the matrix of cooperative principles a mere rhetorical penumbra. The occasional moment of crisis can rekindle a momentary involvement of members. The ultimate vesting of the enterprise in the members--farmers or consumers--rather than a private owner, is of genuine significance. But the difference in actual daily operation is much less than the initial cooperative ideology might suggest.

COLONIAL USE OF COOPERATIVES

This, then, is the model of the cooperative which was extended during the present century to the third world. The initial transplant was under the auspices of colonial governments, and was strongly flavored by the pervasive paternalism of foreign rule. In India the Raiffeisen model of the agricultural credit society appeared to the British Raj as a possible solution to the inextricable problems of rural endebtedness created by the interaction of traditional village hierarchy, Western concepts of property, and commercialization of agriculture.

However, a host of problems limited cooperative effectiveness. The credit cooperatives were frequently captured by the locally dominant caste, and converted into a supplementary institutional resource. Defaults and arrearages ran as high as one-third. The moneylenders were supplemented, but not displaced; they were able to provide credit for purposes, such as ritual needs, that cooperatives could not. They were often less concerned with repayment of principal than with collection of the high interest, and retained means of social coercion within the local community to accomplish this end. Despite all these difficulties, the credit cooperative formula was extended to Sri Lanka by the British, where it enjoyed some modest success. Bruised and battered, but not mortally wounded, the Indian cooperatives survived but did not really prosper in the colonial period.[13]

In French-ruled Africa, the dimension of colonial paternalism was especially pronounced. In many of the territories formerly under French administration, in the postwar period a network of Sociétés de Prévoyance were created as marketing cooperatives. A German cooperative official with many years of work building cooperatives in the developing world, Konrad Engelman, recounts the bitter evaluations of those organisms given by African agricultural officials in a 1962 seminar on cooperative development. The Ivory Coast delegate reported that: "The general mistrust of government aid, specifically the application of cooperative methods, is based on the experience with the administration of the former Sociétés de Prévoyance, the leaders of which were men capable of everything and good for nothing." The Guinea delegate added that: "The unique purpose of the Sociétés de Prévoyance was to collect money without any regard for the farmers' interests. The consequent mistrust prevented until 1957 the majority of the Parliament from consenting to any encouragement of cooperative work by the state." These sentiments were echoed by other participants, who complained of the compulsory membership, enforced contributions, and exclusion of Africans from administrative functions in these colonial cooperatives.[14]

In Africa, the retention of the original Western model of cooperation was greatest in areas of British rule. In Ghana, Tanganyika, and Uganda, cooperatives emerged in the postwar era as a quite significant phenomenon. Although in the 1950s there was active administrative encouragement, participation was never accomplished by administrative fiat. The enabling legislation which accorded legal status to marketing cooperative societies made provision for extensive administrative supervision, but an important zone of autonomy remained to the cooperative. The role of African initiative in the Alliance of Cooperatives in Ghana, the Kilimanjaro Native Cooperative Union, the Bukoba Cooperative Union and the (Sukumaland) Victoria Federation of Cooperative Unions in Tanganyika was considerable. It may be noted that cooperatives emerged at this point exclusively around the marketing of key export cash crops: cocoa in Ghana, coffee in Bukoba and Kilimanjaro, cotton in Sukumaland.

Some particular features of British administrative tradition and colonial history which permitted a higher order of spontaneity in cooperative development may be noted. A less authoritarian style of administration, especially where agricultural policy was not constrained by settler pressures, is part of the explanation. The periods of Labour Party rule in 1929-30 and 1945-51 also had an unmistakable impact in this field; cooperatives were particularly congenial to British socialists. Lord Passfield (Sidney Webb) served as Colonial Secretary in the Ramsey MacDonald government, and gave strong encouragement in the early phases of cooperatives in Bugisu (Uganda) and Kilimanjaro; in Tanganyika, the first Cooperative Societies Ordinance dates from 1932, resulting from initiatives begun with Lord Passfield's support. The postwar Labour government encouraged the adoption of cooperative legislation in many African territories, and established cooperative departments to foster the movement. A number of the British staff recruited for the cooperative service brought to their labors the ideological precepts of the Rochdale pioneers.

Such, then, was the initial transfer of cooperatives to Africa and Asia, passed through the filter of colonialism. Its orbit of operation was rural, linked to the marketing of a narrow range of export crops, or rural credit. A tradition of administrative paternalism had grown with the cooperatives, in sharp contrast to their original modus operandi in Western set-

tings. In fact, the movement had taken root in only a few areas;
the totally administrative cooperatives of Belgian and French-
ruled Africa collapsed immediately upon the withdrawal of their
colonial support.

It is critical to note that the cooperative model which was
introduced was alien. While in some instances political leaders
asserted a correspondence between the cooperative concept and
indigenous social practices and orientations, most notably in
Tanzania, the institutional model actually introduced was Western. There was not in fact a close similarity between the marketing (or producer) cooperative and the kinship-based patterns
of social exchange and reciprocity which form part of the African
cultural heritage. The only recorded instance of significant
cooperative forms which were entirely self-generated and founded
upon local cultural norms was in Liberia, where the state made
no significant effort to promote marketing cooperatives.[15]

We turn in the next chapter to a theoretical exploration of
the potential and limitation of cooperatives. At a deductive
level, we will examine the attractions and problems of cooperatives for the state and for smallholders. We hope in particular
to illuminate the critical issue of equality and efficiency.

2.   Cooperatives and Development Policy

INTRODUCTION

In many of the new nations of Asia and Africa, agricultural cooperation serves as an important instrument of development policy. It is used by government in dealing with the political and economic situation of the rural sector.[1]   The promotion of rural cooperation which frequently accompanied independence was usually founded on a large element of faith in the efficacy of this device.   This is not surprising: there are, as we shall argue, cogent reasons for backing cooperatives, and a number of the associated difficulties become evident only over time.   Indeed, when policy commitments were being made, the bulk of the extant literature concerning cooperatives was the product of writers long immersed in the cooperative movement, totally dedicated to its ethos, and wholly convinced of its moral and practical value.[2]

In the last decade, researchers have attempted to grapple with the lessons of the post-independence cooperative experience. A number of interesting and valuable case studies have concentrated on individual cooperative societies, and on cooperative unions as subsystems in which important social, economic, and political changes affecting society as a whole may be seen at work.   In other studies researchers have used the government's approach to cooperatives as a description and analysis of national political organization and processes.   In a third class of works researchers have applied theoretical principles to explain the causes of failure or success in cooperative operations.[3]

Lacking in the treatment of cooperation and cooperative policies is an attempt to build a generalized framework for analyz-
ing cooperatives, based upon propositions taken from theories of
change and development, within which this particular aspect of
economic development policy can be understood. In this chapter,
we will suggest a theoretical framework, distilled from the accumulated literature and comparative experience.

We first ask why in certain instances farmers and governments try to achieve their goals through cooperative organizations, rather than through other, alternative organizational
forms. Farmer motives for cooperation are the basis for cooperatives as an autonomous social force; government, national political leadership motives help us to understand the nature of government support for and control of cooperation. Of course, the
farmer goals and government objectives are interdependent; and
cooperation is a meeting ground of these forces. Thus, the tension between cooperative autonomy and government control is one
of the recurrent themes of our analysis.

A second recurrent theme presented in this chapter is our
attempt to evaluate cooperation's potential as a means of achieving equality and efficiency objectives.[4] A wide range of social, economic, and political factors affects the ability of cooperatives to contribute to the realization of efficiency and
equality. In order to provide our discussion with a manageable
focus, we concentrate primarily upon the implications of the
motives of the major actors involved in cooperation for the
achievement of these two development goals.

A DEFINITION OF "COOPERATIVE" AND ITS THEORETICAL SIGNIFICANCE

Many types of organization are today referred to as cooperatives. A necessary first step in this analysis, then, is to
identify characteristics and criteria of the cooperative. The
following characterization may be suggested:

> A "cooperative" is a formal organization embracing more
> than one agricultural production unit (household), mem-
> bership in which requires that individual production
> units limit their operational independence, either by
> making use of the organization's services in certain as-
> pects of their operation, or by accepting the organiza-
> tion's intervention in managerial decisions concerning
> the conduct of their agricultural activities.

Cooperative "membership" implies both contribution of equity
capital to the organization and possession of formally equal
rights to participate in democratic processes of organizational
decision-making, without regard to differences in the amounts of

equity capital contributed by the various members. In the polar
case, cooperatives may require their membership to subordinate
their formerly independent units to the cooperative's all-encom-
passing central direction. The legal and operational identity
of former units is dissolved within the "higher stage" cooperative, which may be termed a "collective." However, retention of
the concept of membership, as the definition of the status of the
former farm operators, leaves such collectives within the boundaries of the category "cooperative," distinguishing them from
state farms, on the one hand, and commercial mergers of farm
units through purchase or the formation of multi-farm companies
or corporations, on the other.[5]

But a crucial distinction must be drawn between institutional forms of cooperation, which rely upon external organizational models, and indigenous cooperation, or the extant patterns
of reciprocal exchange and sharing which are embedded in local
rural cultures. The institutional mode of cooperation in the
third world generally traces its pedigree to the corpus of cooperative theory and practice which developed in the West. Further, institutional cooperation invariably hinges upon the incorporation of local production into national and international market structures. The quintessence of indigenous cooperation was
its immediate local focus. Despite the occasional invocation of
superficial similarities between the two modes of cooperation by
promoters of the cooperative creed as warrant for this policy
approach, the differences in their constitutive principles are
quite fundamental.

The common themes of indigenous cooperation are given cogent
summation by John Bennett:

> The indigenous forms of cooperation may vary, but they
> share one important trait: they are all manifestations of
> a functioning social system. They emerge from the nat-
> ural social groupings of society: kinship, ritual segmen-
> tation, age groups, status and class, fraternal associa-
> tions, and mutual-aid organizations. Indigenous coopera-
> tion is part of the activities that societies initiate
> in order to survive and to perform the tasks necessary
> for day-to-day operation. Cooperative action may assign
> equal status to participants, or it may be controlled by
> prevailing hierarchical patterns. Shares may be distrib-
> uted more or less equally, or there may be unequal divi-
> sion or even exploitation of particular classes of par-
> ticipants. When the groups are small, there is a ten-
> dency toward equality; as they grow larger, inequality,
> and the exercise of force and social sanction to compel
> participation, become more common.[6]

While a fruitful graft of the two patterns may in some cir-

cumstances be possible, this does not automatically occur.   Indeed, indigenous concepts of cooperation may well impede extension of the institutional form.   Only infrequently have state
policy-makers taken much account of indigenous patterns in de-
signing cooperative institutions.
    Structurally, the modern, technologically advanced society
is marked by a high degree of differentiation within large-scale
units of social, economic, and political action.   The complex
task of coordinating the many activities of many individuals acting to successfully perform social, economic, and political functions cannot be left to spontaneous interpersonal understandings
and agreement; coordination must rely on self-conscious, rational
administration of activities within and by formal organizations,
whether these be private or public.   The basic challenge lies in
building an organization upon a foundation of individuals willing
and able to fill the administrative and support roles essential
to the organization.[7]
    Thus, with regard to the economy, Johnston states:

> a crucial--perhaps the crucial--requirement for economic
> growth is for the citizenry of a developing country to
> acquire the ability "to concert their individual behav-
> iors into a national (international) network of increas-
> ingly large-scale specialized units of collective action
> which are necessary for development and widespread use
> of increasingly productive technologies."[8]

Brewster lists five types of essential "intermediate" organizations, linking "farm villages and the outside world," which must
be created to promote the economic transformation of the agricultural sector:

    1) supply organizations;
    2) marketing organizations;
    3) educational service organizations;
    4) operational service organizations; and
    5) credit organizations.

In addition, Brewster emphasizes the essential role played by
"progress-oriented stable governments," which both initiate and
direct socioeconomic change and provide a climate of stability
encouraging nongovernmental change initiatives.[9]
    In this last point, Brewster touches briefly upon a central
principle elaborated by Huntington: the critical role of political institutions in building and maintaining the political, social, and economic order in large and complex national societies.
The political system fills a major role in enabling, or in some
instances compelling, individuals and groups to enter new, ex-
panding, and intensifying networks of relations.   As new social

forces are mobilized and brought into contact with one another, special forms of organization are needed to mediate the increasingly important relations between these forces and the authoritative institutions of the state.[10]

The cooperative, as we have defined it, may then be viewed first of all as one of the organizational responses to the demands of the "development and widespread use of increasingly productive technologies." Secondly, the cooperative functions as one of the new intermediary bodies through which government and farmers interact. The analysis which follows will focus upon both of these aspects, their interrelationships, and their implications for the realization of equality and efficiency objectives through cooperation.

FARMER PARTICIPATION IN AGRICULTURAL COOPERATIVES: MOTIVES AND MODES

Cooperation in the underdeveloped nations is typically a movement of small farmers. In the first part of this section, therefore, we consider how cooperation assists small farmers in their relations with the rest of society. We then turn to a more detailed discussion of farmer motives and attitudes toward cooperation, distinguishing between typical small farmers and more wealthy and powerful rural leaders.

Third world cooperative movements depend quite heavily on government backing and often are subject to considerable governmental intervention and direction. While recognizing that this situation limits the influence and importance of farmers' autonomous objectives, nonetheless farmer motives weigh heavily on the factors shaping cooperative operations. We shall not, therefore, exaggerate the real impact of governmental backing and oversight on cooperative operations. Third world governments are in most cases quite weak--indeed, it will be argued below that government choice of cooperatives as one of its policy instruments is a recognition of its policy-making and administrative capabilities.[11]

With the exception of relatively efficient authoritarian regimes, such as those ruled by Communist parties, cooperatives generally are not mere appendages of the state administrative apparatus. Where cooperatives have become large, complex organizations, government is likely to be quite simply incapable of closely supervising and directing the affairs of all cooperative bodies. Government may have nominal control of a number of major cooperative organizations and their leadership; it may control the movement's investment program through its control of credit; it may determine a most basic variable affecting cooperative operations through its fixing of crop prices--but within the broad parameters fixed by central government's national policy decisions, functioning cooperative organizations will continue, under most circumstances, to constitute relatively autonomous arenas. In particular, rural elite elements, who are most likely to achieve substantial personal material gains, tend to devote energies to cooperative affairs, thereby achieving real control of the organization.

Thus, in Uganda, even after government intervention intensified in the period from 1966 onwards, most cooperative unions were not placed under the direct management control of the Cooperative Department; and with only rare exceptions the approximately 2,000 primary cooperative societies continued to function with a considerable degree of independence. Barring special circumstances which might draw particular interventions, or financial crises which necessarily drew an unusually high degree of administrative attention to the affected cooperatives' affairs, a large number of decisions were left to the cooperatives to be made in keeping with their own internal constellations of interest and power.[12] At the same time, we must not lose sight of the fact that cooperative structures themselves are frequently weak, and only imperfectly fulfill these functions.

Small Farmers Within National Society

The small farmer often finds himself at a disadvantage in his economic and socio-cultural relations with the urban center and the commercial intermediaries through whom village-city economic relations are channeled.[13] The cooperative can serve as a means for farmers to unite to strengthen their hand in economic dealings, both as buyers and as sellers. It may also serve to bridge cultural gaps and improve farmers' self-image and their social standing, by demonstrating their ability to deal competently with their own economic affairs, and to serve broader regional or national interests.

Many analysts of agricultural production have noted that the family farm--the farm operated predominantly through the labor of members of the kinship unit whose income is directly dependent upon net farm income--is the most efficient means of organizing the production of many crops. Only a few tropical crops have real economies of scale or capital requirements that give a clear advantage to plantation agriculture.[14] Moreover, historical experience shows that even where economic conditions are unfavorable to the small farm, the family farm shows strong staying power. Subsistence production most often provides the minimal basis required for the continued physical existence of the farm family, even under deteriorating conditions; and the farmer's attachment to land and village very often makes him willing to continue his agricultural activities, even in the face of declining real income.[15]

However, small farmers have a marked disability to direct and influence change, and to compete with the urban populace for

control of scarce resources. The service and manufacturing sectors, by contrast, consolidate at a much more rapid pace, facilitated by technological advances in production, communication, and
transportation. These sectors rapidly enlarge individual production units and coordinate activities within corporate structures
which reach markets of far greater geographical scope than those
of the traditional economy. Thus, the effects of technological
change create a situation in which many farmers face a limited
number of buyers of their crops. Similarly, large numbers of
small-scale farmers come to face a small number of firms supplying the consumer goods, essential services, and physical inputs
to the agricultural production process.[16] This consolidation
process is telescoped into the space of a few decades in the poor
nations of the third world. Public and private sector investment
in nonagricultural economic activities is based upon the import
of modern, large-scale systems of production from the advanced
economies of the world. This means handing a great deal of market power to a few non-agricultural economic groups, since the
limited local markets of the poor nations cannot absorb the output of more than a few large-scale units.[17]

The disadvantages suffered by the small farmer when facing
the superior market power of urban-based, and urban-oriented buyers and suppliers may be compounded by state intervention. The
economic power and organizational advantages of large firms can
be mobilized in the political arena to win favorable policy measures. Instances in which international and local expatriate
firms handling export crops have been able to gain government
backing for cartelization agreements is an example.[18]

The cultural gap which often exists between the average ag-
riculturalist and the merchants and commercial agents who represent the urban center and its particular skills and values further compound the effects of the small farmer's economic and political weakness. However shrewd, the small farmer, often illit-
erate, lacks the contact with, understanding of, and power over
regional and national economic and political organizations which
would help him to conduct his economic relations with commercial
intermediaries on an equal footing. The farmer is easy prey to
unscrupulous practices, ranging from sharp bargaining to outright
fraud, without being able to seek and find influential supporters
to help him defend his interests.

Moreover, even where small farmers are not the victims of
concentrations of market power or the sharp dealings of commercial intermediaries, they may well tend to ascribe the economic
difficulties they encounter to the operation of such factors.
The farmer, suffering the effects of often unfavorable price
movements which result from world market conditions, seeks an
explanation for his distress in factors lying nearer to hand.
Commercial intermediaries are natural scapegoats. They are usually seen as greedy parasites, and where, as is often the case,

traders belong to clearly identifiable minority communities, suspicions of unscrupulous, exploitative manipulations are intensified.[19]

Finally, it must be stressed that the socio-cultural gap between the small farmer and commercial intermediaries and the city has a far broader significance than its immediate impact upon farmer-trader economic exchanges. The clash between village and urban center is perceived as a confrontation of values and ways of life as well. Small farmers often suffer from negative stereotyping. Their lives are depicted as being marked by cultural barrenness and by an economic and technical backwardness which not only holds down their own standard of living, but also constitutes a drag on the entire process of national economic-technological advance.[20]

The Cooperative as Solution

The argument for cooperatives, first of all, asserts their potential for helping the small farmer to overcome these economic weaknesses.[21] The cooperative buys and sells in larger quantities, centralizing the dealings of farmers in the hands of those with greater commercial skills than most farmers themselves. The cooperative thus claims to equalize to some degree at least the farmers' economic power and bargaining skills in relation to suppliers of goods and services and the buyers of farm produce.

Secondly, the formal cooperative may also be a source of "solidary" incentives as well--a means of improving the small farmers' social image and raising their status, in their own eyes and in the eyes of the rest of society. Moreover, successful economic operations may also be valued for their "purposive" aspect, enabling common agriculturalists to feel that they are contributing selflessly to the achievement of the goals of any of several reference groups with which they may identify.[22]

The development of economic grievances and the growth of farmers' awareness of themselves as members of a distinct occupational and/or ethnic group have also contributed to the attractiveness of cooperatives to small farmers. The egalitarian, democratic membership principle of cooperative organization renders the cooperative's structure symbolic of the shared identity and interest of farmer-members as they face the culturally distinct and economically hostile outside world. If some farmers come to see the cooperative as a source of solidary and purposive incentives, the cooperative will enjoy a degree of affective loyalty which may be of great importance in helping it to survive the economic difficulties of the early period of its existence. However, since longer-term cooperative success depends upon economic effectiveness, and since the economic margin available to most farmers is quite narrow, there will be clearly defined limits to

the period during which the cooperative may count on member loyalty to aid in overcoming economic troubles. Whether this
breathing space will be exploited to establish and maintain the
cooperative as a viable institution will depend largely upon the
interaction of the cooperative's rank and file members with the
organization's leaders.

The Typical Small Farmer and the Village Leadership Elite

Thus far we have discussed farmer motives associated with
agricultural cooperation within the context of a consideration
of farmers' position as a socioeconomic group in relation to and,
to a great extent, in conflict with the rest of society. It is
now necessary to turn to a consideration of the various motives
for supporting the cooperatives. To simplify analysis, a sche-
matic contrast will be drawn between two broadly defined categories of farmers: the small farmer and the village leadership
elite.[23]

The intensification of ties to national and international
markets and the diffusion of new productive technologies often
disadvantages the typical small farm just as it awards advantages
to larger, entrepreneurial farmers. Both the real income and
the social standing of the smallholder family may be threatened,
first of all, by a steady process of socioeconomic differentiation within the village community. This process creates a minority of bigger, more technically advanced farms. In addition,
the typical small-scale unit may be squeezed by direct (legal or
extra-legal measures aimed at taking small operators' land) or
indirect (monopolization of access to commercial and government
agencies; depression of market prices through expanded, more efficient production) pressures exerted by more successful farms
and estates.[24]

Prior to the shift in the rural economy from locally ex-
changed production to production for metropolitan markets, both
economic-technical and social factors helped to set fairly strict
limits to the degree of inequality which could arise among the
family farm units within the village farm community. Social
pressures and the structural arrangements of the economy within
such small, largely isolated agricultural villages also tended
to set limits to the accumulation of wealth through large-scale
agricultural activities. Whether the village was dominated by a
large-scale landowner or peopled by independent freeholders, the
community as a collectivity imposed redistributive demands upon
its successful members. Through such measures, the community
helped to guarantee at least subsistence to all its members; it
acted in this manner to prevent violent outbreaks which might
result if shifts in wealth and control of access to the means of
production were not moderated and the physical existence of community members were to be threatened as a result.[25]

In return for his compliance to community redistributive demands, the successful farmer was offered a degree of status and social and political influence within the village community. Facing the numerical superiority backing village pressures, and lacking ties to supra-village economic or political structures which could be activated effectively to provide backing in struggles against the village majority, the wealthy farmer had little alternative but to accept the exchange of wealth and income for social standing and power within the confines of the local community.[26]

The spread of marketing relations and the more effective "penetration" of rural areas by regional and national economic and political structures change this situation considerably. New technologies become available; marketing of produce is facilitated; and new and attractive consumer goods become readily accessible. Wealth could be accumulated in monetary form. Outside elements could form their strongest ties with the more successful and influential agriculturalists; by virtue of their attitudes, skills, and social and material resources, these could offer more rewarding relationships to the commercial agent or government representative.

Such new outside contacts provide the ambitious and capable farmer with the backing needed to burst the social obligations which the village majority had formerly been able to impose upon the scale of farm enterprise. Thus, while a few members of the village community are freed to respond to new incentives and exploit the new opportunities afforded by closer integration in regional or wider markets, the weakening of redistributive mechanisms leaves the mass of small farmers in a steadily worsening competitive position. Moreover, the mass of farmers may also be forced into competition with commercially oriented owners of large estates, who, like members of the village elite, enjoy access to regional and national centers of economic and political power.

The typical farmer suffers a number of disadvantages as he attempts to hold his own in these confrontations. He lacks financial reserves and his income is too low to facilitate innovation. The small scale of his operation makes him a poor credit risk and raises the administrative overhead costs of providing him with credit; thus it is difficult for him to obtain loans for innovation at reasonable rates of interest. Moreover, lack of land makes certain innovating investments unprofitable, even if funds could be obtained. A greater part of his energy and labor resources must be devoted to assuring family food supply. Also, the small scale of his production means that on purely economic grounds, the typical farmer cannot expect to receive the same consideration which the larger-scale producer receives in the market, whether as buyer of production inputs (discounts for bulk purchase; priority in allocation of scarce commodities) or as seller of produce (advance purchase contracts at guaranteed prices; buyer participation in provision of transport). Similarly, those involved in formulating and implementing government's agricultural development policy assume that economic-technical considerations demand that preferential treatment be given to outstanding agriculturalists, to the detriment of the typical small farmer.[27]   Finally, personal favors and gifts, which wealthy farmers are able to provide to commercial and government agents, reinforce job-related economic and technical considerations which work against the typical small farmer.[28]

The small farmer lacks social or political resources to counterbalance his economic standing in relations with the field representatives of business and government. Nor does the mass of small farmers possess organized political power which could be mobilized to demand national agricultural policies aimed at removing the small farmer's disabilities.[29]

Organization offers a possible means of overcoming the difficulties facing the typical small farmer through the pooling of economic-technical, social, and political resources. The theoretical economic and technical advantages of cooperation are several, and hypothetically open the way to both improved efficiency and equality:

1) Establishment of cooperatives can help farmers to overcome the financial barrier to innovation, through the pooling of funds for mutual lending, and by serving as an intermediary between the small farmer and government agencies or private banks. The cooperative makes lending to small farmers possible by taking upon itself the effort and expense of administering many small loans and through its commitment of institutional income and assets to the repayment of the sum borrowed.[30]

2) Where innovation is hampered by small-scale operations which make investment in new equipment uneconomic, cooperation can coordinate members' farm operations so as to make possible sharing of equipment by a number of production units whose total extent is sufficient to justify the investment. Joint tube wells for irrigation are a well known example of such activity, drawn from the agricultural development experience of Bangladesh.[31]

3) In the market, pooling of purchases and sales makes it possible for typical small farmers to achieve a number of commercial gains: for example, the achievement of bulk buying discounts in the purchase of supplies, on the one hand, and ability to submit tenders in competition for special supply contracts, on the other.

4) By aiding small farmers in all these ways to increase their ability to innovate, and by providing a framework within which contacts with large groups of small farmers can be efficiently organized, cooperation makes the possibility of achieving program objectives through work with the mass of small farmers more attractive to government extension agents.

Village Elites and Cooperative Leadership

The cooperatives are usually led by members of the village elite. We have already noted the presence in the village of individuals who, using outside backing to resist village redistributive pressures, are able to exploit widening access to external markets and sources of technical innovations so as to modernize and expand their farms. The more successful farmers generally function as major leadership figures within the village community. They often have the inclination and the means to play such a role. Smaller farmers will often recognize that only the larger cultivators have the commercial skills to lead cooperatives, and thus will demand that they do so.[32] By fulfilling such duties of leadership as the hosting and financing of ritual functions, the extension of material aid to fellows in need, or providing backing in disputes, the better-off become influential patrons, to whom many are tied in informal clientage. Accordingly, the bigger, wealthier farmers are best able to compete for elective positions and to fill the role of cooperative officers.

In the typical case of a marketing cooperative, a number of advantages motivate members of the village elite to seek to lead cooperative organizations. First of all, though they do not own the businesses they direct, cooperative officers can usually derive considerable material benefits from their posts. In addition to receiving direct cash payments, such as allowances and bonuses paid in successful crop years, officers are in a position to control cooperative operations and resources so as to further their own private ends. Officers can, for example, influence matters such as: the hiring of workers; the administration of credit; the acceptance/rejection and grading of produce; and the choice of suppliers of goods and services to the cooperative. Moreover, leaders are often in a position to benefit from acts of outright corruption--embezzlement and influence peddling--with little danger of punishment.

Although cooperative officers' control over cooperative assets and income is somewhat limited by comparison to the private trader's control over his business, the risks borne by the cooperative leader are also relatively smaller than those of the private trader. Various forms of government support for cooperatives may be available to an undertaking organized as a cooperative, which would not be furnished to a private venture. Also, the cooperative officer may be able to mobilize a considerable percentage of needed capital from the farmer-patrons of cooperative services, rather than providing capital personally. The officer does not bear personal financial liability for society losses, as would be the case were he acting as a private businessman. Finally, where minority ethnic group, "pariah" entrepreneurs dominate marketing, processing, and moneylending, enterprising members of the village elite may sense that cooperatives' democratic character makes them the best means of mobilizing communal sentiment behind attempts to dislodge ethnic outsiders from their economic strongholds.

In some cases, village leaders may also be drawn to cooperative office by nonmaterial incentives. Far more easily than private trade, cooperative leadership can be presented to self and to the local community and outside observers as service to the community, thus making possible the achievement of purposive and solidary-status gains while filling an economically profitable role. Indeed, a certain degree of negative social evaluation may adhere to private commercial undertakings, which does not apply to cooperative leadership. The cooperative officer's position as the chosen leader and representative of farmers seeking commonly held goals may then seem far more palatable and honorable than the role of trader, both to members of the village leadership elite and to those observing them, from within the community and from without.[33]

Efficiency and Elite Control

Solidary attachment to the cooperative as a means of closing the gap between the small farmer and the estate operator or of displacing minority group merchants grants the cooperative a "credit" of time in which farmers may voluntarily continue their association with the cooperative despite technical and economic difficulties. However, this respite must be exploited quickly to guarantee the material returns the small farmer expects if mass defection from the cooperative is to be avoided.

Cooperative officers drawn from the village elite doubtless have some incentive to conduct cooperative affairs efficiently, so as to provide economic benefits to all members and thus sustain their loyalty and support. However, experience suggests that many leaders will prefer the short-run advantages of manipulation of cooperative resources for personal ends to the attempt to succeed in managing cooperative affairs according to objective economic and technical criteria. There is, first of all, no guarantee that efforts to manage efficiently will be crowned with success. Moreover, even where efficient management brings marginal gains of several percent to all society members, the gratitude thus earned may be relatively diffuse. By contrast, the misuse of society resources to grant special favors may have a far more significant impact on the welfare of the small number of individuals benefiting from them; and such gains are clearly a product of the specific, personal relation between the society officer and his beneficiary. Thus, special favors may be more effective in increasing the officer's effective influence with the village community as well as securing loyalty to the cooperative for key clients.[34]

Whether officers will reach for immediate gains through manipulation of cooperative resources, or will take the less certain, long-run approach and strive to set cooperative operations on a sound business footing, will depend largely on the pressures which rank and file members are able to bring to bear. Farmers may press officers most readily where alternatives to cooperatives are available, or where cooperative functions are of marginal importance to incomes and can be dispensed with completely without undue hardship. In such cases, the departure of dissatisfied farmers, one by one, from the cooperative's ranks will be sufficient to deplete cooperative resources and ultimately bring the demise of cooperative organization. Where the farm income interests of the officers depend on cooperative services (as in the case of dairy cooperatives, to be discussed below), the possibility of cooperative liquidation will be doubly threatening. Disbanding cooperatives in such cases means not only officers' loss of control over cooperative resources, but also serious losses in their farm income.

Sustained pressure within cooperative structures in order to achieve reform and improvement is more difficult for the mass of farmers. The need to organize overt, joint action is far more difficult than simply quitting since it exposes supporters to retribution. Since typical small farmers tend to lack the social and economic independence and the organizational skills required to exercise effective, critical oversight of officers' behavior, the theoretical importance of government's role as promoter, guarantor, and arbiter of cooperation among the rural populace is clear.

Equality and Elite Control

In considering the possible egalitarian role of agricultural cooperatives in third world countries, it must be emphasized that the redistributive pressures formerly applied by the village community were an outgrowth of the cultural milieu and the functional demands of village solidarity, rather than the product of a concept of equality itself as a basic moral principle and social goal. This contrasts sharply with ideological formulations of the principle of equality, argued by Coleman to be central to the "development syndrome."[35] Radical egalitarian creeds, such as contemporary socialist ideology, depict the poor as the bearers of a new moral order and emphasize equality as a major ethical principle and social objective. In most rural societies, however, the astute, skillful, and thrifty husbandman who expands his wealth and income may be envied, feared, and made the object of community pressures, but he is at the same time seen as a man worthy of respect, epitomizing community virtues and individual capability. In short, the successful farmer and the distinction he has achieved, rather than some abstract concept of social equality, represent the ideal to which the members of the tradi-

tional farming community aspire. The very worst off in the community may be guaranteed certain assistance through community
redistributive mechanisms, but are nonetheless viewed as unsuccessful. The community aids them without idealizing them.[36]
Thus, neither the village elite nor the mass of small farmers
are likely to provide an intravillage egalitarian ideological
impulse to agricultural cooperatives.[37] Of course, where
cooperative organization encompasses members of more than one
community (whether defined in religious, racial, ethnic, or other
terms), the inherent weakness of the egalitarian impulse will be
compounded by intercommunal tensions, which may be exploited by
members of the local elite in their struggles to control cooperative resources.

It seems reasonable to conclude, then, that the building of
cooperatives which will work actively toward the reduction of
intra-village inequality cannot be based solely upon intra-village forces. Cooperatives must receive the backing of supra-
village structures seeking to achieve similar goals and possessing adequate resources to defend typical small farmers acting
within the cooperative framework against retaliation by the village elite. At the same time, efficiency may be a prerequisite
to equality, in that only through the material benefits thus
arising can smallholder loyalty be secured.

GOVERNMENT COOPERATIVE POLICY:
CHOICE OF THE COOPERATIVE AS A POLICY INSTRUMENT

In previous sections we noted the importance of governmental
policy as a factor helping to shape the evolution of cooperation
in the underdeveloped countries. Indeed, in most instances governmental support, and even coercion, have been the main factors
accelerating the spread of cooperation and raising the level of
membership to a significant percentage of the rural population.
In this section, then, we will attempt to identify the factors
which may lead the state to promote cooperation as a vehicle for
intervention in rural society and economic activity.

Intervention in the agricultural economy is widely seen as
an imperative by the political elites of underdeveloped nations. The words of a Uganda Government white paper may be seen
as representative on this point: "[agricultural] production and
marketing . . . are political issues and [Government] must
therefore be associated with them."[38] Given the predominantly
agricultural basis of most third world economies, state intervention in agriculture is necessary, in the first instance, so
as to provide government with vital local and foreign currency
revenues. In addition, since agriculture provides employment
and incomes to a major proportion of the population, intervention
is required to allow national political leadership to reward

supporters, punish opponents, and prevent the growth of autonomous centers of economic, and thus political, power within the
rural sector. Finally, intervention is readily justified in
terms of widely held beliefs concerning its efficacy--indeed,
necessity--as a means of promoting economic growth.[39]

The cooperative is an ideal mechanism for intervention. It
transfers agricultural operations of a multitude of individual
small farmers to the framework of a formal body, registered and
recorded in government files, and operating within the context
of government laws and regulation. Thus, cooperation transforms
innumerable, anonymous, informal, and unrecorded activities and
transactions into clusters of relatively parallel and uniform
behaviors and exchanges, channeled through the cooperative and
recorded by it.

Government objectives may be achieved, however, by means
other than cooperation. The question thus arises: Why should
governments choose to be "associated with" agricultural production and marketing through cooperatives, rather than adopting
the two major alternative approaches:

1) governmental regulation of private economic activity,
   or
2) direct governmental conduct of economic activity
   through the agency of government ministries or governmentally owned parastatals?

Historical and Technical Background of Policy Choice

Historical and economic-technical realities of agriculture
in the developing countries influence the leadership's decision
to use cooperatives or choose other means of intervention. Where
a new regime inherits a large cooperative network and a governmental agency skilled in dealing with cooperatives, this will
pull, other things being equal, the government toward favoring
cooperation. This factor has been of great importance in shaping
government policy in a number of Asian and African states where
the colonial power had built strong cooperative structures.

Political considerations, however, may lead national policy-
makers to forego the advantages of working through established
organizations. This will occur where the cooperative leadership
is viewed as ideologically and politically hostile, or merely
suspect. Thus, for example, we argue in Chapter 9 that the contrast between early post-independence policy toward cooperatives
in Ghana and Uganda may be explained in terms of political leaders' evaluation of the political positions of the heads of the
cooperative movement: government hostility in the Nkrumah years
in Ghana, and support most of the time in Uganda.

Cooperation in most instances must also be reasonably efficient and technically rational to assure political backing over

any extended period of time. Where the economic and technical demands of production or marketing do not require and reward greater concentration and coordination, cooperation is an arbitrary vehicle, incapable of providing real material gain and hence maintaining farmer enthusiasm. In such a situation cooperation will either break down of its own accord; or it may become a transparent, and dispensable, facade for government control.

Thus, government efforts to promote cooperation, and those of farmers as well, often concentrate on crops and operations which require and reward increasingly larger operations. Thus, the crops examined in our study--cocoa, cotton, and coffee--all must be bulked in large quantities for shipment to foreign markets. In addition, cotton and coffee must be prepared for export by primary processing carried out in relatively large and capital-intensive plants. Milk, a domestic consumption product whose cooperative handling is also discussed below, shares these characteristics: bulking and cooling are needed to make possible sale of milk to distant, domestic urban markets. Economies of scale, therefore, bring their own reward.

By contrast, food crops which are purchased unprocessed by the final consumer are not a fertile ground for cooperation, since they can easily be disposed of in small quantities in a multitude of local markets. Thus, for example, although a formal cooperative monopoly in the handling of various local food crops was declared in Uganda in 1970, no serious implementation attempts were made. The declaration of monopoly over food crops was quickly cancelled after the change of regime; on the other hand, cooperative control of the more amenable cotton and dry-processed coffee was left in effect.[40]

Cooperation versus the Regulation of Private Economic Activity

Two economic activities often influence policy-makers to prefer promotion of cooperation to the regulation of private economic activity: production, and marketing and processing. The advantages of cooperation as a medium for governmental attempts to direct farm production activity lie, quite simply, in the ability of group work methods to multiply the impact of restricted numbers of government field staff. Through contact with the cooperative, government efforts to promote innovations will reach greater numbers of farmers than will individual farmer contacts. Moreover, since the cooperative provides a framework within which typical small farmers may work together to overcome shared economic and technical disadvantages, cooperative organization not only enables greater numbers of farmers to hear of suggested new approaches to production but also places more farmers in the position of being able to respond to such promptings.

As for marketing and processing of agricultural produce, the choice of cooperatives rather than regulation is a function of government desire to control but without threatening other entrepreneurial groups in the private sector. Private enterprise is, even in the poor countries, a fairly well established socioeconomic institution with legal rights, a high degree of autonomy, and privacy in the conduct of its activities.[41]

The belief that cooperatives are more readily accessible to government control than private undertakings was expressed quite openly by various Uganda People's Congress (UPC) Ministers of Agriculture in Uganda. These politicians preferred the cooperatives over African private companies in Africanization programs because of the government's greater authority and ability to intervene in cooperative affairs.[42] However, no parliamentary or legal barrier stood in the way of using the UPC majority in the National Assembly to amend the Companies Act to expand the state's powers of intervention. Fear of the economic and political disturbances which might result from a change in the status of the private economic enterprises which dominated Uganda's economic life may well explain why the UPC government did not try to alter this law, but to emphasize cooperatives instead.

A further advantage of cooperative methods in the field of marketing and processing is that, in countries where minority group entrepreneurs dominate agricultural marketing and processing, political leaders use farmer resentment as a source of backing for their regimes. They find the cooperative an attractive means of transferring economic control to local hands. Through cooperative membership large numbers of citizens become directly and personally involved as beneficiaries of government localization programs, within organizational frameworks which maintain close, continuing relations with supervising governmental agencies.

Cooperation versus Direct Governmental Operations

Governments in less developed countries find it difficult to control the agricultural sector. Control is hampered, first of all, by the problems of mobilizing additional funds and personnel. In addition, direct responsibility for the conduct of agricultural activities places heavy burdens upon the policymaking system and strains the professional capabilities and disciplinary mechanisms of administrative agencies.[43] Given such problems, policy-makers cannot equate greater government involvement in agriculture with greater effective control in their own hands. Awareness of these limitations is another reason why policy-makers consider cooperatives as an alternative to direct control. Regulated cooperatives help extend governmental influence, while lightening the burdens imposed upon the state's material resources and its policy-making and administrative capabilities.

By involving farmers themselves in the conduct of activities

which government seeks to direct, cooperatives mobilize new, un-
tapped resources: personnel (both elected officers and cooperative employees) and financial (members' shares, deposits, and
loans). The cooperative, rather than the government, absorbs
losses incurred in cooperative operations. Cooperation can also
serve as a means of decentralizing some portion of agricultural
decision-making, thus lessening the strain upon central decision-
making bodies. Furthermore, if left a reasonable degree of autonomy, cooperatives may come to serve as channels of information, providing independent reporting on agricultural conditions
and the performance of government field staff. Since most underdeveloped countries lack effective agricultural interest groups
or a strong agricultural or local general press, such independent
feedback is typically lacking. This factor, which hinders both
policy-making and effective professional and disciplinary supervision of government field workers, can be ameliorated by cooperatives' input of information.[44]
    Finally, the cooperative can provide a buffer of local, non-
governmental, leaders who absorb the initial impact of farmer
discontent caused by instances of mismanagement or by national
economic conditions which harm farm incomes. National political
elites are aware of the weakness of the tools with which they
grapple with complex agricultural questions, as well as limitations in the professional competence and sometimes probity of
local level field staff. State elites thus are likely to appreciate the high probability of encountering setbacks in their attempts to guide the agricultural sector. Therefore, governments
may be willing to allow local partners to share in the credit
for successes, so as to ensure the presence of local, nongovernmental leaders who will bear the immediate onus of blame in moments of failure. In sum, if one takes into account the fact
that political leaders often have little grounds to believe that
direct governmental operations would be any more efficient or
controllable, one can understand why a number of third world governments have persisted in attempting to make cooperation work
in the face of chronic weaknesses in cooperative functioning.
    Political and social values may also play a role in encour-
aging the choice of cooperation. Third world leaders with socialist preferences may be drawn to the cooperative by its potential contribution to equality. The cooperative can, in theory,
serve as a means of achieving an egalitarian distribution of
trade and processing profits in programs designed to eliminate
commercial middlemen and place farmers themselves in control of
such activities.
    In addition, given its democratic membership structure, co-
operation has been seen as achieving egalitarian goals through
means which involve the mass of small farmers in national development efforts and promote civic virtue within the framework of
the state. In the words of a senior Ugandan Co-operative Department official:

> The cooperatives have provided some education to its
> [sic.] members in management and honesty and trained
> some members to become public spirited. It has aroused
> the interest of ordinary man to work hard and also real-
> ise that he can also play a role in the economic devel-
> opment of his country.[45]

Government Choice and the Issue of Efficiency

In many instances, pressing difficulties make the gradual
strengthening of cooperative membership oversight of the organization impractical and demand immediate intervention by the appropriate state agencies to correct dishonest or incompetent management practices. The probability of government intervention
aimed at guaranteeing efficient performance is, however, limited
by a number of factors.

Government interest in deriving immediate political benefit
from its support for cooperatives is a major consideration which
may restrain efforts to promote efficiency. Sound management,
if it can be achieved, brings moderate gains to all members and
produces good feeling which may in the longer run be convertible
into active political backing. On the other hand, provision of
relatively large, selective benefits to critical village leaders
offers the prospect of winning immediate, tangible support from
these leaders and their followings. Use of cooperation to
achieve such an end requires that government look the other way
as village elites exploit cooperative office as a further means
of enriching themselves and cementing their dominance.

Consideration for the democratic and voluntary nature of
farmer involvement in cooperatives may also restrain government
intervention. Respect for democratic values is at stake here,
but more practical considerations are also involved. Through too
heavy-handed a policy of intervention, genuine farmer interest
and identification with cooperatives will extinguish, and with
it the advantages listed above in our discussion of cooperation
and direct governmental conduct of agricultural operations.

Finally, and most simply, the same factors which make it
difficult for government to control agricultural operations,
make it difficult for government to guarantee cooperative efficiency. Supervision and preventive, as opposed to remedial, intervention are likely to be beyond governmental capabilities for
the majority of cooperative organizations. Government will,
therefore, have to confine itself to resolving disputes and eas-
ing crises.

Government Choice and the Issue of Equality

The degree to which governmental cooperative policy serves
to promote equality will depend, we suggest, upon three variables: the ideological orientation of national political leader-

ship; the socioeconomic basis of support for the national political leadership; and the leadership's longer-term political and
economic strategy. The implications of the first two principles
for equality are rather straightforward: policy and its administration are more likely to be directed toward the promotion of
equality to the extent that national leaders are normatively
committed to egalitarianism and to the extent that their political power is based on the backing of groups which stand to benefit from the promotion of equality.

As for the third factor, the promotion of equality through
cooperatives demands a well-calibrated, long-term strategy. We
have noted above the tendency of socioeconomically dominant individuals within the village to assume control of leadership positions within cooperatives. Where national political leaders feel
compelled to win the backing of village elites, the cooperative
comes to serve as a meeting point through which benefits are provided to dominant village figures in exchange for their continued
support for the regime.

Although acceptance of the village elite's controlling position in cooperatives may serve short-term goals of maintaining
support for the regime, longer-term strategy may well point in
the direction of attempting to raise a new cadre of cooperative
leaders from among the mass of typical farmers. Such leaders are
dependent first of all upon the backing of party and governmental
agencies, and in addition upon the support which they are able
to mobilize among their peers with the assistance of these supra-
village political bodies. Here an attempt is made to circumvent
the leadership elements of the existing socioeconomic structures
and to reach the masses through new leadership drawn from their
ranks, and hence more responsive to egalitarian goals. Such a
approach is certainly harder, for it entails the building of
novel structures on the basis of new leadership and an egalitarian mass appeal, and will probably arouse the opposition of the
advantaged strata of rural society. However, for the political
elite with the means and the political vision to undertake it,
such a strategy offers the prospect of building a firm local political base for their national leadership role.

The chapters which follow present our research findings on
Uganda and Ghana. These chapters are grounded on the analytic
framework presented here, as a means of organizing and deepening
our understanding of the potential of cooperation as an instrument of development policy in the third world nations.

3. The Political Economy of Uganda: An Overview

ESTABLISHMENT OF BRITISH RULE

In this chapter, by way of introduction to our presentation of the cooperative experience in Uganda, we present a brief historical overview of selected aspects of the Uganda political economy. No effort is made at a comprehensive résumé of Ugandan political history; an extensive literature is available on this topic.[1] Our purpose here is to offer enough historical background to provide a contextual setting for our analysis.

Britain functioned as imperial midwife for both Uganda and Ghana. In the Uganda case, missionary action preceded formal colonial expansion, with the Church Missionary Society (CMS) represented at the court of the King (Kabaka) of Buganda from 1878. A decade later, in 1888, the Imperial British East African Company was chartered, with a mandate to establish a mercantile-political zone of influence inward from the coast. By this time, highly competitive mission action in Buganda--with the CMS battling French-supported Catholic evangelization and Muslim supporters fostered by Zanzibari traders--had brought the Kingdom of Buganda to a state of civil war. Company agent Frederic (Lord) Lugard tipped the balance in favor of the Protestant faction, which paved the way, in 1893, for the signature of a treaty of "protection" between Buganda and the British government. The East African Company dropped from the picture, and in 1894 London confirmed its "protectorate." With the aid of Buganda allies, the territorial domain of the Protectorate was extended far beyond the initial confines of Buganda itself, to encompass the current boundaries of Uganda.

Over the decade which followed annexation, the colonial state took form, based upon an extraordinary imperial legion composed of a handful of British officers, a small Indian detachment, loosely disciplined irregulars of Sudanese origin, and, most numerous, the Ganda levies of the Protestant faction. During the first three decades of colonial rule Ganda agents were widely employed in chiefly positions in most districts. Also, an unusual treaty was negotiated with Buganda in 1900, conceding special status to the kingdom, and awarding land titles to 4,000 of the leading members of the new ruling class.[2] Buganda was also the base camp for Christian expansion; Ganda catechists did much of the field labor as dedicated extension personnel for the missionary endeavor.

Buganda was pivotal in the new colonial economy. The thin ranks of the colonial administration were under considerable pressure to generate local revenues to cover their costs. Though the promoters of annexation had claimed in 1893 that $20,000 a year would meet the costs, by 1899, $397,000 were required and no major revenue source was in sight. By 1901, the Uganda Railway had reached the shores of Lake Victoria at Kisumu, and some traffic to begin to amortize its costs was also needed. Cotton came to the rescue.

In 1903, experiments began with imported cotton seed. The following year, chiefs in Buganda were instructed to have their peasant dependents plant cotton plots along accessible roads; a year later the first 241 bales were exported. By 1908, a single variety of seed had been selected and generalized, and the beginnings of a buying and ginning infrastructure were taking shape. Levies on the cotton cash flow began to provide the state with a reliable revenue base. The foundations of a cotton-based peasant economy were laid.

In contrast to the cocoa situation in Ghana, the colonial state here had played a decisive role selecting and distributing the seed, and inducing the Buganda chiefs to require that it be grown. The chiefs in turn pursued cotton imposition with considerable enthusiasm, perhaps because they perceived that a commercial crop would permit higher rents from the tenants on the far-flung estates they had acquired under the 1900 agreement. However, direct coercion to require cotton planting lasted only a short time, and there was an important element of peasant receptivity in the rapid spread of the crop. As Wrigley has argued, "The truth seems to be that the response to economic opportunity, though by no means wholly spontaneous, was rather more spontaneous than in other parts of East Africa."[3]

Promotion of cotton cultivation as a central preoccupation of the state thus became an institutionalized habit, and has persisted ever since. Pressure for expansion came from three sources: (1) the urgent necessity to generate bulk traffic to meet capital charges on the East African railway; (2) the need to generate revenue for the state itself, by imposing a crop permitting peasants to pay their head tax in cash; and (3) a strong interest in cheap and reliable imperial cotton supply by British textile interests. Further, cotton required a capital commitment to a processing industry, supplied by immigrant interests. Once the ginneries were built, their profitability depended largely on volume processed, thus institutionalizing a powerful expatriate lobby within the colonial society pushing for state promotion of peasant production.

Table 3.1

EARLY DEVELOPMENT OF COTTON PRODUCTION

Year	Number of Bales (400 lbs.)	Value in £
1905/6	241	1,089
1907/8	3,973	51,594
1909/10	6,209	59,596
1911/12	13,378	165,412

SOURCE: C.C. Wrigley, Crops and Wealth in Uganda (Kampala: East African Institute of Social Research, 1958), p. 15.

By World War I, cotton cultivation had spread into the eastern districts of Busoga and Teso. Government revenue expanded from 62,000 pounds in 1900 to 283,000 pounds in 1914; cotton by that time provided 60 percent of the export earnings. The modest British Exchequer subsidy, which amounted to a grand total of 2.5 million pounds from 1893 on, came to an end. The new colonial state could now extract enough local resources to lubricate its machinery.[4]

INTERWAR COLONIALISM IN UGANDA

The interwar years in Uganda saw the consolidation of the imperial state and the colonial economy. The colonizer was now strong enough to shed the alliance with Buganda sub-imperialism, and confirm his direct hegemony over all parts of the country, including Buganda. Coffee emerged to take its place beside cotton as the productive base of the colonial economy. The racial

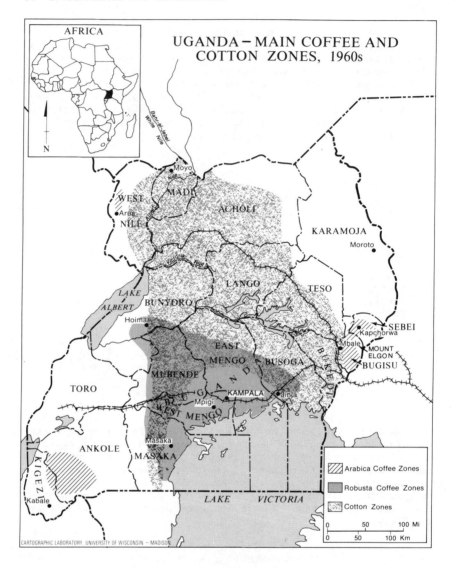

division of labor of economy and society became clearly demar-
cated, with African growers, Asian marketing and processing intermediaries, and European exporters. Political consciousness
in Buganda began to spread beyond the mission-educated elite to
the peasant mass.

In Buganda, British officers began to exercise direct tutelage over the regional administration. New forces were stirring
in the Kingdom: the clan heads (Bataka), who believed themselves
heavily disadvantaged in the 1900 settlement, articulated their
own discontent, and simultaneously served as a vehicle for peasant grievances over growing exactions of the chiefly allies of
the colonizer. A new generation of colonial officers was sympathetic to these grievances, and at the same time made use of them
to impose reforms on Buganda which enhanced the predominance of
the colonial state, as well as mollifying rural discontents.

The land issue had first been posed by the Bataka movement
in the early 1920s, which claimed that many clan estates had, in
effect, been expropriated in the land distribution which followed
the 1900 agreement between Britain and Buganda. Although lands
linked to the 40-odd clans themselves had generally not been dis-
turbed, estates tied to ceremonial places of origin and burial
grounds of the much larger number of lineages and sublineages
into which clans subdivided had frequently been allocated as estates. In its first form, the dispute ended as a draw, with the
British ruling that the Bataka grievance was substantially justified, but that it was impractical to overturn the 1900 distribution of mailo land. (Mailo, a deformation of the English word
mile, denoting the unit of measure widely used in the distribution, came to describe the 8,000 square miles actually allocated
as official estates, which were attached to public office in the
Kingdom, or private holdings.) By 1926, attention had shifted
to the status of tenants on mailo land; the chiefly beneficiaries
at this point derived their revenue primarily from levies imposed
upon tenants, rather than capitalist exploitation of their holdings. With the spread of cotton, the traditional levies could
be substantially increased, both through the cash settlement required in lieu of the month of unremunerated labor in the propri-
etor's fields (busulu), and in the tithe (envujo) imposed on
crops produced. Before cotton, envujo obligations were settled
through a delivery of some beer and bananas; a government committee claimed that by 1926 landlords were demanding as much as one-
third of tenant cotton.[5]

Under British pressure, the Buganda Lukiko (Council) reluc-
tantly passed in 1927 the busulu and envujo law, which had far-
reaching effects. The law established a fixed, 10 shilling rate
in lieu of labor service, and, in effect, established envujo payments at 4 shillings per three-acre plot. Equally important, it
removed from the landlords their rights of summary eviction of
tenants, according to the latter quasi-permanent security of ten-

ure. Although the prescribed payments still yielded attractive
revenues to landlords at the time the legislation was adopted,
the rates were never altered, and over time the payment became
increasingly nominal.[6]

As prologue to the emergence of coffee as an African peasant
crop, we may note the brief encounter with European plantation
agriculture. In the years just prior to World War I, Europeans
had acquired 58,000 acres in 135 plantations; of this, 104 plantations, and 50,000 acres were in Buganda, all mailo land sold
with the consent of the Lukiko. About 10,000 acres were planted
in coffee by war's end, and half as much in rubber. After a
short-lived price boom in 1919, markets for both products turned
sour. The colonial administration, at this juncture, was not
prepared to engage in forcible labor recruitment to assure a
servile work force for the plantation operators; Liverpool and
Manchester in any case were pushing for directing peasant labor
to cotton. Plantation agriculture thus withered on the vine.
Some holdings were subsequently purchased by Asians, who ultimately enjoyed some success with tea and sugar. But agricultural
production became fundamentally a peasant enterprise.[7]

Robusta coffee, found wild in Uganda, was ideally suited to
the soil and rainfall ecology of the Lake Victoria region. The
busulu and envujo law, which offered sufficient security of tenure to make perennial cultivation a worthwhile risk, unlocked
the door. Coffee acreage in African hands rose from 959 acres
in 1925, to 4,673 in 1928, and 16,970 in 1931. In Buganda, the
new crop received very little official support; indeed, the lin-
gering plantation interests vigorously opposed African entry into
this sector. However, the labor requirement for coffee was substantially less than for cotton, and returns in most years were
very much greater. As with cocoa in Ghana, the establishment of
robusta coffee in Uganda is largely attributable to African initiative. Even as the trend of a shift into coffee was gathering
force, it remained only dimly visible at first; as skilled an
observer as Lucy Mair, writing in 1934, claimed that the peasant
was inflexibly committed to cotton, and would not switch to another crop even if assured of profitability.[8]

The government role was greater in establishing the arabica
coffee industry in Bugisu and Sebei, on Mount Elgon. Arabica, a
higher value product, grows well between 5,000 and 7,000 feet,
and the volcanic soils and relief rainfall of the mountain assured favorable growing conditions. The first seedlings were
distributed in 1912, and the first 11 tons exported in 1915. By
1922, there were still only 400 acres planted; by 1928, this had
slowly expanded to 1,134, and then began to grow more rapidly to
reach 7,375 in 1938. Output grew from 50 tons in 1925 to 4,000
tons in 1940. By the 1920s, growing numbers of young Gisu were
returning from periods of labor on European farms in Kenya, where
they had become familiar with techiques of coffee cultivation.[9]

On the cotton side, there was both a huge increase in the size of the crop, and its rapid spread to the eastern and northern provinces; however, by the end of the 1930s a production plateau had been reached. The 1938 record crop of 424,000 bales was not equaled again until 1965. Seed was distributed free, and government provided standardized planting instructions through the chiefly bureaucracy. Production increases were obtained almost solely through expansion of acreage; the only noteworthy innovation was the use of ox-plowing in Teso district, where there was a long pastoral tradition. Substantial administrative pressure was exerted in the form of "propaganda," and indirect economic constraint was exerted through the generalization of a head tax, which had been initiated in Buganda in 1900, but fully applied in all areas only after World War I. But there was no direct coercion to secure an expansion from 47,000 bales in 1920 to the 424,000 bale record.

Cotton growing for most cultivators was superimposed on household food production. With hand tools and family labor, it was impossible to work more than a couple of acres, nor could the individual cultivator hope to acquire more than a small income from cotton. Substantial profits could accrue only through extraction of small surpluses from large numbers of planters. As will be noted below, this was possible in the downstream phases of cotton ginning and exporting. For a time, in the 1920s, some chiefs were able to parlay claims to "traditional" tribute labor into sizable revenues. In addition to the Buganda case already noted, this was especially true in Busoga, where chiefs could require from their subjects one month of unpaid labor on fields which attached to the office, or a cash payment of 10 shillings. With the introduction of cotton, the official fields became, for a brief decade until the system was abolished in 1925, official plantations for the private benefit of the chief. Fallers records that the ruler of the largest county of Busoga had an income of over 3,500 pounds in 1925.[10]

Both cotton and coffee required processing before export which was beyond the scope of the peasant producer; around these functions grew the racial division of labor. Cotton had to have its seed removed through ginning. Coffee cherry had to have its outer skin removed, the inner skin around each of the two beans eliminated, and be thoroughly dried. By World War II, all the cotton, and 80 percent of the coffee acreage was African. Asians predominated as buyers and, to be lesser degree, in processing, while European firms played the dominant, though not the exclusive role in export sales.

During this period, the government increasingly entered the picture to arbitrate the sharpening struggle between British and Asian commercial interests, and to assure a price level to the grower high enough to guarantee a reasonable supply; its own survival depended on revenues now generated primarily by export

Table 3.2

COFFEE ACREAGE, INTERWAR PERIOD

	1922	1925	1928	1931	1934	1936	1938

Non-African Planters

	20,820	18,884	18,408	17,559	13,391	13,472	13,314

African Planters

	1922	1925	1928	1931	1934	1936	1938
Buganda	600	959	4,673	16,970	21,050	27,570	32,255
Bugisu	400	475	1,134	2,114	4,023	6,080	7,375
Western province	0	353	2,924	1,816	5,276	9,853	12,238
Total	1,000	1,787	8,731	20,900	30,349	43,503	51,868

SOURCE: Uganda Government, Report of the Committee of Inquiry into the Coffee Industry (Entebbe: Government Printer, 1967), pp. 1-2.

Table 3.3

COTTON PRODUCTION BY PROVINCE, 1929-1939
(000 bales)

	1929	1931	1935	1938
Buganda	82	82	131	179
Eastern (Busoga, Bukedi, Teso, Bugisu)	94	105	81	192
Northern (Lango, Acholi, Bunyoro, West Nile)	26	16	36	51
Western (Toro)	-	-	1	3

SOURCE: Uganda Government, Report of the Uganda Cotton Commission 1938 (Entebbe: Government Printer, 1939), p. 119.

Table 3.4

COTTON AND COFFEE PRICES AND PRODUCTION, 1929-1938

| | COTTON | | COFFEE | |
Year	Average Price to Grower (sh/100 lbs.)	Production (000 bales)	Average Price f.o.b. Mombasa (sh/100 lbs.)	Exports (metric tons)
1929	17	203	86	1,864
1930	15	130	63	2,227
1931	11	190	46	3,182
1932	11	203	51	3,955
1933	9	291	42	4,545
1934	10	278	38	7,000
1935	12	249	37	5,727
1936	10	322	33	10,409
1937	13	331	32	11,727
1938	8	424	23	12,727

SOURCE: Uganda Government, Report of the Uganda Cotton Commission 1938 (Entebbe: Government Printer, 1938), p. 119.

taxes on cotton and coffee. The "ruinous competition" so often
bemoaned in official reports of these years tended to raise
prices to growers. The elimination of buying competition by the
state to protect the interests of British capital then required
the further step of fixing prices to growers. Otherwise the
oligopoly structure created by the state might so depress the
buying price that the supply was threatened. Step by step, cotton and coffee became totally regulated industries.
    Government first moved in 1918 to ban hand ginneries, which
had become quite popular in the Buganda countryside, on the
grounds of difficulties of quality control. This effectively
sealed the farmers into the primary production sector. The next
target were the "middlemen," small Asian entrepreneurs who in
the early years combed the countryside with a pair of scales and
a few hundred shillings capital; their competitive buying clearly
gave higher prices to farmers, but squeezed profits at the ginning and exporting level. During the 1920s, the independent
middlemen were effectively legislated out of operation, with
ginneries now employing their own buying agents.
    Until 1914, ginneries were exclusively European. Asian
competitors then entered the field, and during the 1920s there
was a rapid rise in the number of ginneries, increasing to 176

in 1926, 85 percent of them Asian. In that year, an embargo was
placed on new ginnery construction. By 1929, ginning pools began to form; in 1932, administrative allocation of the market
through designation of specific buying zones became policy. As
Mamdani concludes, "An industry that was almost solely European
owned and managed at the time of World War I was just as completely Indian owned and managed by World War II."[11]

The pattern set in the cotton industry was swiftly applied
to coffee. In 1930, an ordinance required that all African robusta be graded for export; the graders were attached to the 12
curing works which belonged to three major agricultural trading
companies, Old East Africa Trading Co., Bauman, and Jamal Ramji
and Co. In 1936, the legislation became more restrictive, giving exclusive licenses for buying and curing to the three major
firms.[12]

THE TORTUOUS POLITICAL ECONOMY OF TERMINAL COLONIALISM

In Uganda, the pathway to independence was tortuous. The
role of the immigrant communities, a sharply defined district
focus to African societies, an intensely felt religious division
--all these were emotion-laden issues. Looming over all else
was the vexed question of the status of Buganda. Armed with the
1900 treaty, self-assured in its sense of superiority, accorded
a deference by the colonizer denied to all other districts, advantaged by every measurable indicator, Buganda found it impossible to contemplate the modest, one-man-one-vote arithmetic
promised to its 15 percent of the population.

At the close of World War II, Buganda alone had a politically mobilized rural population. Serious disorders occurred in
1945 and 1949; beyond the immediate issues that triggered them,
the riots reflected generalized grievances over agricultural
prices and Asian domination of marketing. The first modern political party, the Uganda National Congress (UNC), was established in 1952, heavily Ganda in leadership, and initially
largely built upon discontent which at that juncture was articulate only in Buganda.[13] The UNC did endeavor to spread a nationalist message to other districts; though it did begin to
have some impact, its missionary efforts were soon eclipsed by a
dramatic confrontation between Buganda and the colonizer.

This crisis, which was to deflect the movement toward self-
rule for several years, was provoked by the utterly different
conceptions of political evolution held by British officials and
the Buganda kingdom. Sir Andrew Cohen arrived in 1952, with a
decolonization mandate and a strong personal commitment to the
unitary state formula. This Buganda adamantly opposed. When
Buganda refused to participate in the Cohen blueprint, he responded by deporting the Kabaka (king), and was faced with a
suddenly galvanized Buganda, united in indignation at this af-

front. Two years of impasse could be resolved only by bringing back the Kabaka in 1955, with major concessions to Buganda in the form of further autonomy.

Thereafter, Buganda sporadically boycotted the central institutions, and even declared its separate independence in December 1960, a gesture which had no effect. Buganda did not participate in the first elections for the Legislative Council African seats in 1958. General national elections were first organized for a fully elective National Assembly in 1961; this time Buganda authorities ordered a boycott of the election, which limited participation in the Kingdom to only 3 percent. Finally, with independence on the horizon, a Buganda-centered political movement, Kabaka Yekka (KY), emerged to win a sweeping triumph within Buganda. The Kingdom finally gave grudging acceptance to a national government, after extracting major concessions: KY participation in the central government as coalition partner, designation of the Kabaka as Chief-of-State, and above all consolidation of the autonomous status of Buganda.

The intensity of religious cleavage was another impediment to national consensus, dating from the religious factionalism in Buganda in the nineteenth century. The sharpness of religious rivalry was not equally strong in all regions; it was most tangible in West Nile, Acholi, Kigezi, Ankole, and Buganda. However, it became reflected in political party competition. In 1956, with the close informal backing of the Catholic Church, the Democratic Party (DP) was launched; from the outset, it attracted a largely Catholic following. The Uganda People's Congress (UPC), established in 1960, was identified in many eyes as predominantly Protestant.[14]

Thus, as independence drew near, there were three major political movements: UPC, DP, and KY. A first set of national elections was held in 1961. With boycott orders from the Buganda royal leaders, the results were perverse: the UPC won the most votes, but the DP held the parliamentary majority, 41-35, on the basis of Buganda seats garnered with the ballots of the 3 percent who braved the boycott to go to the polls. (KY was formed after the 1961 elections.) On this basis, the DP formed a short-lived government in 1961-62, under Benedicto Kiwanuka. In the decisive 1962 elections, the UPC won a clear verdict over the DP outside Buganda, winning 37 seats to 24. In Buganda, KY took all 21.

The independence formula was determined by the coalition calculus of hostilities. The UPC and KY had one common trait: their ferocious animosity to the DP. This provided a short-lived basis for coalition and was the formula for independence. The UPC was forced to accept a semi-federal constitution, and to pledge its support for the designation of Kabaka Mutesa II as ceremonial Chief-of-State. This paved the way for independence on 9 October 1962, under Prime Minister Obote.

In the economic sphere, substantial growth occurred during

this period, but the underlying cotton-coffee structure was little changed.   The exceptional prices in the early postwar years permitted rapid expansion of the state and its socioeconomic infrastructure of schools, health facilities, and roads.   After 1955, prices were much less favorable, and revenues failed to keep pace with the momentum of state-enlargement and social development which had been set in motion.   Government expenditure multiplied eightfold during the period, from the very modest level of 4.5 million pounds in 1947 to 32.3 million pounds in 1962; as a fraction of GDP, they nearly doubled.

Some diversification of the economy did occur during these years.   The Owen Falls dam at the Jinja outlet of Lake Victoria was begun in 1954, reaching by 1962 a capacity of 150,000 kilowatt hours (nearly half of which was sold to Kenya).   A number of light industries grew up in the Jinja-Kampala area, producing textiles, beverages, food products, and other consumer goods.

Table 3.5

GROWTH OF THE UGANDA ECONOMY, 1946-1962

Year	Monetary GDP	Exports	Government Expenditure (million pounds)
1946	21.4	11.6	–
1947	24.6	13.6	4.5
1948	30.3	17.1	6.5
1949	42.8	26.6	6.7
1950	54.3	32.9	8.0
1951	83.8	51.5	12.4
1952	88.3	51.0	16.0
1953	76.3	40.4	17.4
1954	92.8	48.1	19.1
1955	102.0	49.8	22.0
1956	102.8	44.9	23.5
1957	109.4	51.2	24.8
1958	105.9	51.6	25.6
1959	108.0	47.3	25.3
1960	110.8	48.3	25.8
1961	111.2	46.1	28.7
1962	107.9	44.7	32.3

SOURCE: Dharam P. Ghai, _Taxation for Development: A Case Study of Uganda_ (Nairobi: East African Publishing House, 1966), p. 17.

Table 3.6

EXPORT STRUCTURE, 1950-1960

Year	Cotton	Coffee	Total (million pounds)
1950	16.7	8.3	28.7
1951	28.7	13.7	47.2
1952	29.9	12.3	47.2
1953	16.8	11.5	33.4
1954	20.9	13.5	40.6
1955	16.4	20.1	41.9
1956	19.3	15.7	40.4
1957	17.5	21.6	45.9
1958	18.1	20.8	45.9
1959	15.4	18.7	42.1
1960	14.9	17.0	41.6

SOURCE: International Bank for Reconstruction and Development, The Economic Development of Uganda (Baltimore: Johns Hopkins University Press, 1962), p. 444.

The railway was extended west to near the Zaire border, permitting the opening of a small copper mine at Kilembe (Toro). But coffee and cotton continued to provide over three-quarters of the exports throughout the period.

More closely examined, beneath the apparant stability, important changes were occurring in the production of both crops. Though total cotton acreage remained fairly stable, averaging about 1.5 million acres, there was a pronounced shift in the regional distributions. In Buganda, increasingly, cotton was a crop grown only by immigrant tenants without secure tenure, while Ganda switched massively into coffee. The loss of cotton acreage in Buganda was offset by the steady expansion in the area cultivated in the north, which had produced only small amounts before World War II, and also in Busoga.

The reason for the shift is quite simple: where both crops could be grown, coffee was almost always more profitable. The coffee acre needed only 80-100 man-days of labor, while cotton required 122-140. An acre of coffee produces about 1,180 pounds of dried cherry (kiboko), while cotton yields, under the most favorable estimates, were no more than 500 pounds per acre.[15] In addition, coffee could be interplanted with food crops more

readily. As Table 3.7 demonstrates, in most years after 1950
the coffee price to growers was significantly higher per pound
than that for cotton. If one were to take an average of the
various yields and labor requirement estimates, and assume an
equivalent price of 50 East African cents per pound, the return
to the critical resource of labor would be 6.56 East African
shillings per man-day in coffee, as opposed to 1.54 for cotton.
Peasant producers were perfectly capable of making this calculation: while total cotton production stagnated, African coffee
production rose from 27,000 tons in 1946 to 114,000 tons in 1962.
    The state was slow to perceive the shift, and did little to
encourage it. Arabica coffee production was given support in

Table 3.7

COTTON AND COFFEE PRICES TO AFRICAN GROWERS, 1946-1962

Year	Seed Cotton	Robusta Coffee (kiboko) (East African cents/pound)
1940	11	6
1946	18	15
1947	20	17
1948	22	19
1949	30	21
1950	33	25
1951	45	40
1952	50	50
1953	50	70
1954	51	150
1955	61	75
1956	55	75
1957	56	80
1958	58	80
1959	47	62
1960	48	52
1961	55	50
1962	57	55

SOURCE: J.J. Oloya, Coffee, Cotton, Sisal and Tea in
         the East African Economies (Nairobi: East
         African Literature Bureau, 1969), pp. 45, 60;
         J.D. Jameson, ed., Agriculture in Uganda, 2nd
         ed. (London: Oxford University Press, 1970),
         p. 375.

Bugisu, but in volume this was only 5-10 percent of the total.
Until the late 1950s, the agricultural extension apparatus of
the state was heavily tilted toward cotton, and the propaganda
campaigns of chiefly exhortation exclusively so. Local governments received a levy of 1.50 shillings per hundred pounds of
cotton output, which added incentives that percolated down
through the chiefly hierarchy to keep pressure on farmers to
maintain cotton acreages.[16] Joan Vincent, reflecting on the
apparent paradox of maintenance of cotton production in Teso
despite the low returns, concludes that: "the final explanation
can only lie in the outside interests that maintained cotton
cultivation as the mainstay of the Teso economy and girded about
the quality of its production with administrative and local government sanctions. . . . Since cotton was the crop of the paternal administration it occurred to few Gondo parishioners (apart
from those who had been out and about in the world) that they
would do better with other crops."[17]

In this period the state further extended its mercantilist
control over all phases of the cotton and coffee industries.
During the war, the entire cotton and coffee crops were purchased
at fixed prices by emergency marketing bodies. These improvised
bodies became permanent after the war, eventually becoming known
as the Lint Marketing Board and the Coffee Marketing Board. The
former European and Asian export trading houses remained as con-
tractual agents for private overseas clients, while the Boards
directly handled bulk sale arrangements, especially to state
purchasers. Only the Bugisu arabica crop, and the small quantities of immigrant estate coffee, remained outside the marketing
board framework.

Until 1953, the government allowed producer prices to rise
only very slowly, despite the buoyant world prices. Vast surpluses soon began sloshing about in the marketing board coffers,
reaching 38 million pounds for cotton and over 9 million pounds
for coffee. Of this, 25 million pounds were ultimately diverted
to other development projects, thus making this component of the
"stabilization" funds precisely equivalent to a tax.

A heavy toll was also levied on the growers through an export tax on both crops; this had been introduced in 1919 for
cotton, and 1915 for African coffee. Between 1945 and 1960, export levies on coffee and cotton amounted to 68 million pounds,
constituting nearly one-third of government revenues, and as
much as 52 percent in 1952. All told, the export taxes and sums
transferred out of the price stabilization funds amounted to 27
percent of the total income which would otherwise have accrued
to growers. The impact was somewhat heavier on cotton (31 percent) than on coffee (21 percent). If we add to this the cess
at times levied on cotton and coffee, the incidence of import
taxes on goods purchased by peasants, the local graduated (alias
head) tax, and school fees, the fiscal extraction of the state

through diverse levies exceeded 50 percent of the cash income of
many peasants.[18]   Farmers paid much higher tax rates than non-
Africans or African wage-earners through the 1950s and 1960s.[19]

A final, unofficial levy on the farmers, particularly in
cotton, was dishonest weighing at the buying station.   Allegations concerning cheating go back to the 1920s, at a time when
the absence of standard measures made it difficult for the peasant to evaluate and compare prices.[20]   Competition among buyers and common measures appeared to diminish the problem, but
the imposition of buying monopolies in the 1930s gave a new impetus.   The 1938 Cotton Commission had taken official note "of
the widespread existence of cheating and malpractices," and cited
Department of Agriculture findings that "in the majority of the
zones growers were cheated of appreciable percentages on the
weights of cotton delivered for sale."[21]   The 1948 Cotton Commission reiterated these charges in sharper tones: "We have
been greatly impressed and profoundly shocked by the volume and
convincing nature of the evidence which we have heard as to the
widespread deliberate cheating of the grower . . . we are satisfied that in the main the allegations are proved up to the hilt
and that the growers are fully justified in their firm belief
that this form of cheating is rampant."[22]   They estimated that
the farmers lost about 10 percent of the cotton value through
dishonest buying.   With the emergence of competition from cooperatives and the approach of independence this form of fraud appears to have diminished.

While the magnitude of the export duty was little visible
to the farmers, cheating was an immediate issue, and frustration
quickly spread over the pricing policies which were diverting
large sums into the stabilization funds.   The 1948 Cotton Commission noted with surprise the "remarkable unanimity of views"
throughout the country on "the relatively low price which the
grower receives considering the high world price for cotton."[23]   In the distant and isolated Bwamba region, Winter ob-
serves that, "when the stabilization schemes were set up the
government went to considerable lengths to explain their purpose
and operation to growers.   The net result is that the Amba are
very resentful of these schemes, and consider their operation
little better than plain theft."[24]

It is thus hardly surprising that anger ran deep in the
countryside, especially in Buganda where political awareness was
highest.   It is against this background that one can understand
the rapid though short-lived spread of the Uganda Federation of
African Farmers led by Ignatius Musazi (later founder of the UNC)
in 1948-49, the widespread demands that Africans be permitted to
take over the processing of their own crops, and the massive response to the exile of the Kabaka.   From 1952 on, the colonial
administration sought to reduce rural discontent by permitting
prices to rise toward the market level, to actively encourage

the formation of cooperative societies, and to facilitate African entry into cotton ginning and coffee processing through the cooperatives. This latter process we will examine in more detail in a subsequent chapter.

The political economy of colonial capitalism had thus produced by the eve of independence a society with an entrenched racial division of economic function and privilege, and with growing animosities built around these disparities. In the African case, these hostilities were focused with particular force on the Asian sector, with which African growers were most directly in contact. Disparities in ecological endowment--the Lake Victoria zone and Mount Elgon produced the higher value coffee crop--and the prewar treatment of the north and west as labor reserve areas resulted in pronounced regional disparities in per capita incomes. To these were added the keenly felt religious division, a widespread resentment of Buganda privilege, and a complex web of ethnic rivalry and local factionalism.

INDEPENDENT UGANDA: COTTON, COFFEE, OBOTE, AND AMIN

Uganda achieved independence in 1962 with a precariously balanced political formula. Prime Minister Obote showed considerable acumen in his early years in gradually enlarging the scope of his power. In 1962, he was hemmed in by the autonomy of Buganda, the large and vocal parliamentary opposition, the more than ceremonial prerogatives of Mutesa II as Chief-of-State, the substantial powers devolved upon elected district councils, and the European and Asian control of the economy. Eight years later, Uganda had become a unitary, one-party state.

Confrontation with Buganda was probably inevitable; any central government was likely to find intolerable its lack of control over the wealthiest region of the country, in which the capital itself was situated. By late 1964, Obote had enough votes to dispense with the KY coalition. At the same time, Obote opponents within his party, and the Buganda establishment, began to lay plans for his parliamentary ouster. An atmosphere of growing conspiracy provided Obote the occasion in March 1966 to suspend the constitution, and assume full powers. Buganda protested the illegality of the suspension of the constitution, and finally in May 1966 ordered central government offices out of Buganda. Obote then called the army into action, dissolved the Buganda government, and forced the Kabaka into exile. The supremacy of the central government was affirmed, as the state apparatus again assumed the hierarchical form characteristic of the colonial state.[25] In 1969, the advent of the one-party state was proclaimed.

In 1969, Obote endeavored to restore the flagging legitimacy of his regime by buttressing its ideological content in a "Common

Man's Charter." In this move to the left, note was taken of the growing social inequalities: a "new political culture" was to be created, where greed and avarice would disappear in favor of moral incentives. The most important move pursuant to this new, vaguely socialist state doctrine was the announcement in May 1970 that the government was assuming control of 60 percent of the 84 major industries and financial institutions in the country.

However, it soon became apparent that improvisation played a very large part in the move to socialism. No reliable figures on asset value were available to the government, nor was there an alternative to retaining the former operators on a management contract basis, transferring all risks to the government and giving a guaranteed return to the former owners.[26] To these indecisive results was added the evident contradiction between the "new political culture" of elite austerity and moral incentives, and the conspicuous commercialism of many top political figures. Most public appraisal of the move to the left thus ranged from skepticism to downright cynicism.

On the eve of the Amin coup, the Obote government was paradoxically at the peak of its power, yet had fatal weaknesses: 1970 was a year of record economic growth, 11 percent over 1969; export values were twice the 1960 level, and revenues were buoyant; the UPC was stronger than ever before, and even Ganda elites were beginning to participate in growing numbers. Yet an atmosphere of disabused detachment remained, and the hostility in the Ganda countryside remained strong. The silent drainage of capital, noted by a World Bank mission as early as 1961,[27] continued through the 1960s, and certainly accelerated after the move to the left. Between 1966 and 1971, over $100,000,000 of capital slipped out, mostly in 1970, reflecting especially uncertainties in the Asian community. With the monetary GDP totaling $800,000,000 in 1970, the loss was very substantial.[28]

The greatest flaw of all in the Obote power formula was the growing reliance on security forces. A mutiny in 1964, over Africanization and related material issues, was swiftly put down by British troops. However, in contrast to the government reaction to simultaneous mutinies in Tanzania and Kenya, the mutineers were rewarded in Uganda. The army was then placed in the role of decisive arbiter by its deployment in the confrontation with Buganda in 1966. Idi Amin, who had become army commander in 1966, steadily increased his personal control over the bulk of the army garrisons.

By 1970, Obote had come to fear Amin, and began trying to isolate him. Amin riposted by consolidating his own grip on key units. By the time of the coup, the armed forces were honeycombed with faction and intrigue.[29] In January 1971, Amin seized power, almost certainly to pre-empt a move by Obote to eliminate him.[30]

Initially, the Amin regime was well received in many quarters: Buganda, delirious at the spectacle of Obote humbled; many civil servants, distrustful of the growing party role; some intellectuals, who felt their freedom of expression was being circumscribed; immigrant economic interests, delighted to see the leftward move halted; Britain, irritated at Obote's aggressiveness on southern African questions; Israel, which imagined that it had special ties to the new President. The initial joy soon had a bitter aftertaste for all these groups. But beyond the accumulated resentments toward Obote, Amin at first attracted support through his personal attributes: populist in style, earthy in expression, physically imposing, apparently courageous.

Initially, Amin presented his new order as an interim, technocratic regime. The first cabinet was composed of six civil servants, two diplomats, two East African Community functionaries, one professor, and only one officer. Commissions of inquiry set to work documenting the iniquities of the Obote regime, focusing on the notoriously corrupt National Trading Corporation, and the controversial and unpopular Coffee Marketing Board. Majority control of 18 of the 85 nationalized enterprises was returned to the private scotor, and further implementation of the Obote socialist decrees was frozen.

Whether or not Amin ever had any intention of returning power to civilians, it soon became clear that the new regime intended to stay. Although there was no unifying perspective which informed the policy choices of the Amin regime, the impact of its measures was quite radical. The most abrupt and sweeping transformation was the 1972 "final solution" to the Asian role in the economy.[31] In August 1972, Amin announced that Asians would have to leave the country. While at first the full scope of the expulsion was unclear, before long it became evident that the measure applied to the 12,000 citizen Asians as well as approximately 50-60,000 others;[32] further, their property had to be simply abandoned. Some 3,500 enterprises and $400,000,000 of physical property thus became available for patrimonial distribution; many were absorbed into a burgeoning, poorly managed, and deficit-ridden parastatal sector. Deep-seated animosities toward the Asian community, deriving from their economic role and social aloofness, guaranteed immense popularity for this melodramatic move. British interests were progressively taken over as well; without benefit of doctrine, Amin more systematically dismantled the immigrant economic infrastructure than Obote's leftward move ever would have.

The state domain under Amin became progressively militarized. While the first Amin government had only one soldier, by the time of the May 1977 defection of Henry Kyemba there were only two civilians left. Even before the council of ministers became staffed by soldiers, it met irregularly, and its demoralized and terrified members were afraid to speak. At the same time, an apparatus of terror was constructed. On the pretext of repressing armed violence in the countryside (kondoism), Amin established a Public Safety Unit, recruited in his West Nile homeland, and in neighboring southern Sudan. An even more sinister instrument was the State Resesarch Bureau, which was attached to the Presidency. The Military Police were converted into a third lawless instrument of intimidation.[33] The army itself, though racked by repeated internal crises and bloody purges, also played its part in the politics of fear.

Insecurity for Amin and increments of terror interacted. The bitter struggle within the army which accompanied the coup triggered purges of suspected Obote loyalists in the security forces which continued for months afterwards. Obote forces, including many military personnel who fled the purges, regrouped in Tanzania, and in September 1972 launched a bedraggled invasion force into southwestern Uganda.[34] Not only was the force decimated and many of its leaders executed, but a further wave of reprisals followed; it was at this juncture that terror hitherto confined mainly to the struggle for control of the army, was extended to the civilian population on a large scale.[35] In 1974-75, no fewer than eight well-documented coup and assassination attempts occurred; plots continued to surface regularly, with each conspiracy raising the terror level one further notch.

The first impact of the Amin coup was to decapitate the political superstructure of the state, but to leave intact, with considerably enhanced powers, the technocratic civil service. By this time, Uganda had an excellent public service, well-qualified, self-confident, little touched by the suspicions of commercialism and corruption which afflicted some of the political overlayer. In most technical spheres--agriculture, veterinary service, medicine--Amin had little knowledge or interest. Only slowly did the Amin system ooze and seep through the specialized services of the state. As the eddies of capricious, arbitrary, and unpredictable violence and terror spread through state and society, Ugandan elites began to flee, especially those like doctors or professors with internationally transferable credentials. Only after several years did acute shortages of skilled personnel begin to become manifest.

As support for the regime vanished, outside the immediate circle of beneficiaries, its survival increasingly depended on access to sufficient resources to equip and reward its coercive apparatus. In 1971, Amin discovered he could demand cash from the Bank of Uganda for "intelligence missions." By the mid-1970s, Amin financed himself and his state by forced advances from the Central Bank. When coffee prices soared in 1976, Amin began exporting large amounts directly by plane to Djibouti and London, with the planes returning laden with imported commodities for direct distribution to the military ethnocracy.[36] The state was not only praetorian, but a predatory parasite on economy and society.

The best-known statistic concerning the Uganda of Idi Amin
is the estimated number of its victims. The International Commission of Jurists, in a scathing report published in 1977,
estimated the number of assassinations at over 100,000.[37]
Kyemba suggests the figure is 150,000.[38] After the overthrow
of the regime a figure of 300,000 was widely cited.

In the economic sphere, there was reasonable growth during
the Obote years, with the developmental momentum carrying into
the first couple of years of the Amin period. The gangrene which
then progressively spread through the arteries of the state began
to reveal its effects, with negative growth rates appearing in
1973 and 1974. The extraordinary coffee price boom in 1976-77
provided windfall profits for the regime, which gave an artificial appearance of robustness; external reserves rose to nearly
$100 million at the end of 1976, and the value of exports that
year totaled 196,000,000 pounds, compared to only 42,000,000
pounds in 1960. However, whereas coffee was only 41 percent of
the 1960 total, it was over 90 percent in 1976. Although the
Financial Times described this 1977 financial outlook as stronger
than at any time since the coup, when more closely examined these
figures reveal the pathology of the regime.[39] The coffee
trade was the sole enclave of value creation, whose revenues were

Table 3.8

MAJOR UGANDAN EXPORTS, 1950-1977
(000 metric tons)

Year	Coffee	Cotton	Copper	Tea
1950-54 average	37	64.2	0	1
1960	117	60	13	3
1967-69 average	179	71	16	15
1978	80	11	2	11

SOURCE: International Bank for Reconstruction and De-
         velopment, The Economic Development of Uganda
         (Baltimore: Johns Hopkins University Press,
         1962), p. 446; Uganda Government, Uganda's
         Plan III: Third Five-Year Development Plan
         1971/2-1975/6 (Entebbe: Government Printer,
         1972), p. 44; Commonwealth Secretariat, The
         Rehabilitation of the Economy of Uganda,
         2 vols. (London: Commonwealth Secretariat,
         1979), vol. I, p. 2.

largely drained into the predatory patrimonialism of the military
ethnocracy. An historic opportunity to convert a conjunctural
windfall into broad-based economic development and independence
was lost.

While the economic expansion of the Obote years was perhaps
not spectacular, it nonetheless made some significant progress
toward diversification. Copper had steadily expanded. Cultivation of tea, a British estate crop in the 1950s, had expanded
impressively through very successful African peasant schemes.
Tobacco showed promise in Bunyoro, Acholi, and West Nile. Large-
scale capitalist farmers began to emerge in Buganda in the late
1950s, developing new lines of very profitable production in
dairying, vegetables, and fruits and flowers for the European
winter market. The construction industry thrived, and with it
building materials (cement, bricks). There were quite significant gains in the light industrial sector--beverages, foods,
textiles.

The coffee expansion is particularly intriguing, because so
much of it occurred through peasant initiative. From the mid-
1950s, a consistently pessimistic attitude prevailed on price
prospects for coffee, and production was not promoted by government, except in the high-value arabica zone of Mount Elgon. By
1962, Ugandan participation in the international coffee agreement
carried a treaty obligation to block expansion of production.
Not only were new plantings of Buganda robusta forbidden, but it
was declared policy to shift as much of the total quota allocated
out of robusta into the higher priced arabica.[40] Though it
was generally believed that plantings had been discouraged by
price and fiat, the 1970 production figures demonstrate the ineffectiveness of these policies; good weather, and favorable
prices created by the failure of the 1969 Brazilian crop, produced a remarkable harvest of 240,000 tons, or more than twice
the 1960-62 levels. The survey conducted by Richards et al.
among capitalist Ganda large farmers in 1966-67 shows that these
operators were deserting coffee in favor of more profitable lines
of production, such as dairying. Smallholders, however, lacked
the control over production factors to enter these spheres; they
evidently continued to plant coffee, though labor investment in
maintenance of the trees and thoroughness of harvesting fluctuated with the price.[41]

Cotton advanced more slowly, and never reached the long-
proclaimed goal of 500,000 bales. However, the 1938 production
record was finally exceeded in 1965, 1966, 1967, and again in
1970. The limited profitability and high labor requirements of
cotton adequately explain the inability to meet the goals set by
planners; what is more surprising is that production prospered
as well as it did. In the early 1950s, it was assumed that 1.5
million acres represented a real ceiling, and that further advances in output could be obtained only through raising

Table 3.9

COFFEE AND COTTON PRODUCTION, PRICE TO FARMER, 1962-1976

Crop Year Ending	COFFEE Production (000 m. tons)	Price (V.Sh/lb.)	COTTON Production (000 m. tons)	Price (V.Sh/lb.)
1962	119	.60	34.2	.57
1963	158.2	.50	67.2	.57
1964	172.4		72.0	.51
1965	152.1	.50	79.0	.56
1966	153.9	.40	82.2	.60
1967	166.4	.40	85.4	.40
1968	133.0	.40	60.4	.45
1969	247.2	.40	75.2	.50
1970	221.0	.48	84.8	
1971	175.7		74.8	
1972	183.7		74.5	.57
1973	196.4		77.6	.57
1974	184.3		36.0	.61
1975	178.3		35.0	.80
1976	157.8		23.6	.86
1977	156			
1978	80		11	

SOURCE: Uganda Government, Statistical Abstract, 1973; J.D.
Jameson, Agriculture in Uganda, 2nd ed. (London: Oxford
University Press, 1970); J.J. Oloya, Coffee, Cotton,
Sisal and Tea in the East African Economies (Nairobi:
East African Literature Bureau, 1969); Commodity Year-
book 1977; Commonwealth Secretariat, The Rehabilitation
of the Economy of Uganda (London: 1979).

yields.[42] By 1970, although there is little evidence that
yields had significantly increased, there were 1 million additional acres in cultivation.   Indeed, the new extensions of cotton growing were even greater, as cotton planting in the traditional Buganda zones continued to shrink, replaced by acreage in
the north.
     The impact of the Amin regime was felt first in the cotton
sector; planting is an annual decision here.   By 1973 shortages
of farm implements, such as hoes, began to be evident in the
cotton zones.   Inflation began to take hold, and decreasing
quantities of goods such as sugar, salt, and other low-income

consumption items reduced incentives to produce (though taxes and school fees still had to be paid). The 1976 crop had dwindled to 135,000 bales, the smallest since 1930, and acreage was down to about 1 million acres. By the end of the Amin era, output was a mere 50,000 bales.

Declining output soon spread to other spheres. Coffee production began to sink, and an increasing fraction was smuggled into Kenya. Tea estates were abandoned, with only smallholder production continuing. Sugar output dropped from 144,000 tons in 1970 to 12,000 in 1978. Cement production dwindled from 191,000 tons in 1970 to 73,000 in 1978, when it ceased entirely. The copper mines closed in 1978 as well. The pool of trucks dropped from 7,000 to 1,600. Significantly, the sole commodities for which output was maintained were beer, waragi (local gin), and cigarettes.[43]

The Amin years left the Ugandan polity, society, and economy utterly devastated. The full measure of the ruin was taken only after the Tanzanian army ousted the Amin regime in April 1979.[44] Hyper-inflation; shortages of the most basic goods; large unpaid foreign debts; declining GNP; a worthless currency; a heavy burden of deficit-ridden parastatal organizations; an eroded physical infrastructure of roads, vehicles, and buildings; a corroded administration; an essentially black market (magendo) economy: these were the bitter legacy of the Amin years. Mere recuperation from the tyrannous depredations of the Amin era will be prolonged; regaining the reasonable momentum of the 1960s is problematic for the foreseeable future. If read with the proper sense of perverse irony, one can perhaps understand the true meaning of the citation by Makerere University when coerced into granting Amin an honorary doctorate in 1976: "never in history has a leader done so much for his people in so short a time."[45]

4. Uganda Cooperatives and the Political System

THE FIVE PHASES OF COOPERATIVE DEVELOPMENT

We intend in this chapter to locate the Uganda cooperative structures in the broader social and political system. Two dimensions at once suggest themselves. In the first place, in the ongoing process of social competition, mobility, and change, what role do cooperatives play as arenas for conflict over resources, and vehicles of social promotion? Secondly, what identifiable political role do the cooperatives play? Before turning to each of these issues, we need to outline the historical development of the cooperative movement in Uganda.

Five phases of cooperative development are discernible, defined by their changing social and political roles. In a period of embryonic development against many obstacles, cooperatives were a channel for protest leadership. From 1952 to 1962, in a phase of rapid expansion with government encouragement, cooperatives were an important avenue for emergence in local leadership roles, which for many could be converted into high political status subsequently. In a third period, from 1962 to 1966, cooperatives enjoyed strong regime backing, substantial autonomy, and were an important source of free-floating social and economic resources for distribution of patronage and access to capital. With a tightening of government supervision and control from 1967 on, entry to important cooperative roles was increasingly governed by bureaucratic criteria, and passed through official administrative channels; cooperatives came more to resemble ancillary organs of state penetration of the countryside. Finally, during the Amin years, cooperatives were afflicted by the institutional decay which struck all other sectors.

Table 4.1

GROWTH OF THE COOPERATIVE MOVEMENT

Year	Total Membership	Cotton Bales Handled (lint)	Coffee Handled (metric tons)
1950	24,993	4,010	479
1955	117,047	19,823	14,098
1960	211,214	81,114	24,414
1965	450,590	267,420	36,040
1970	560,000+	466,000	200,000*
1978	1,100,000	50,000*	

SOURCE: Uganda Government, Report of the Committee of Inquiry into the Affairs of All Co-operative Unions in Uganda (Entebbe: Government Printer, 1968), p. 17; Commonwealth Secretariat, The Rehabilitation of the Economy of Uganda (London: 1979), p. 55.

  * Estimates.

    In phase one, cooperatives were largely a response to the disadvantageous terms of trade imposed on peasant farmers by the mercantile monopolies of the immigrants.   The first beginnings may be traced back to the establishment of a short-lived growers' association by four Ganda farmers in 1913, only a few years after commercial cotton production was introduced.   In 1923, the Buganda Growers' Association was launched, the lineal ancestor of the Uganda Growers' Cooperative Union established in 1933.   Before World War II, the lack of legal standing, which made impossible acquisition of credit from lending institutions, and the strong hostility of immigrant marketing monopolies, placed insuperable obstacles in the path of its operation.   But it was nonetheless significant as a portent of developing rural consciousness of perceived exploitation.   After the war it became the first large, effective cooperative union, acquiring its first ginnery in 1950.[1]
    This new awareness of the disadvantages created by monopolies of trade also found pre-war voice in a letter from the Kabaka of Buganda to the Governor in September 1929 that, "We do not approve . . . the formation and compelling associations (Sindicates) [sic.] to function and receive the British Government's assistance.   We find . . . that the more competition is in existence the better price the grower gets, and furthermore

instead of giving a correct price the syndicate reduces it; and
we therefore find that it is not necessary for such associations
to exist and ruin cotton industry and also the British Government
should not interfere with this urging them to be formed and receive its support."[2] In 1934, every county chief in Busoga
signed a letter of protest against the operation of the ginning
pools. In 1938, the Mengo District Commission wrote that, "the
existing system of cotton purchases is tolerated by the grower
only because it is the only way in which he can dispose of his
crop. Without exception all growers distrust and dislike the
pools, or syndicates as they are generally called and regard
them as being formed solely to depress the price, and as the
pools arose from government policy so is Government also implicated."[3]

The ephemeral British Labour government of Ramsey Macdonald
(1929-31) brought a flicker of official interest in African cooperatives. On 19 June 1931, Colonial Secretary Lord Passfield
wrote to Governor Sir W.F. Gowers that the object of the Bugisu
Coffee Scheme should lead to "the formation of a cooperative organization among the producers themselves and it should be the
aim, kept steadily in view, to hand over the operation to such
an organization as soon as suitable persons selected from the
participants are forthcoming."[4] In a 1933 article in _Africa_,
which reflected the enlightened paternalism of colonial reformers, Strickland argued that cooperation "creates a new social
fibre, a new integration, new communal restraints, and is of
great value to the community and the race. Cooperation will be
a stabilizing, not a subversive force." The same author went on
to invoke a neo-Lugardian legitimacy for the cooperative formula,
which was held to be the translation to the economic sphere of
the doctrine of indirect rule.[5]

Legislation to confer legal recognition on African cooperatives was crafted in 1935, but came under withering attack by
private European and Asian interests when submitted to the then
wholly non-African Legislative Council in 1937. It was claimed
that cooperatives carried "grave risks" of acquiring political
power and doing "incalculable harm" to immigrant ginning interests, while the "incapacity of the native" was said to ensure
their economic failure. The colonial administration was not prepared to stand up to this onslaught, and offered no institutional
channel for the African voice to be heard; the legislation was
accordingly withdrawn as "premature."[6]

With the return of a Labour government to power in 1945,
colonial policy again became lightly tinctured with Fabianism.
A 1945 Colonial Office dispatch declared that "cooperatives are
the most effective media for introducing to the people of the
under-developed areas the brave new world of commerce." Further
circulars instructed colonial administrations to see to the
adoption of suitable legislation permitting cooperative organi-

zations. Its particular form, however, was shaped by the luke-
warm administrative endorsement of cooperation, and the outright
animosity of immigrant interest groups.   The 1946 cooperative
ordinance reflected the desires of the European and Asian mercantile interests to stress the "controlling and restrictive features," and to minimize official encouragement.   In Engholm's
pungent summation of the ordinance, "All that was required to
strangle the cooperative movement at birth was an unimaginative
and schoolmasterish Registrar [of cooperatives] heading an under-
staffed department, ready to impose a mass of niggardly rules on
any association foolish enough to join."[7] In fact, relatively
few did; by 1952, when new cooperative legislation was brought
forward, there were only 278 registered cooperative societies,
with 14,832 members; by far the most important was the Uganda
Growers' Cooperative Union,[8] officially registered in 1948.
The latter was eclipsed soon after its registration by the mete-
oric rise of the Uganda African Farmers Union, led by Ignatius
Musazi.
     In different guises, the Musazi Union was the primary cata-
lyst of rural discontent in the 1948 to 1952 period. The peasant
malaise surrounding the marketing of both cotton and coffee, as
well as price levels, ran deep in Buganda and was seeping into
other formerly quiescent corners of the Protectorate. The Musazi
organization, which claimed a peak membership of 80,000, and established beachheads in Busoga, Bugisu, Teso, and Lango, beyond
its major stronghold of Buganda, lacked the organizational and
financial resources to make any real dent in marketing; as a protest vehicle, as we shall see in more detail below, its impact
was considerable. With the arrival of a new Governor, Sir Andrew
Cohen, in 1952, the colonial authorities changed course, and
sought to contain the protest impulses within an officially sponsored cooperative movement. The groundwork for this was laid by
a far more accommodating cooperative act in 1952, providing
enough autonomy to make registration acceptable. At the same
time, provision was made for compulsory acquisition of a number
of ginneries to be provided to cooperative unions, with government loans to facilitate the financing. With cooperative control
of Bugisu coffee also foreshadowed, and provision for both eliminating discriminatory price policies and offering private African access to Buganda coffee processing, the framework for the
second phase of rapid cooperative development was set.
     In the terminal colonial decade, 1952-62, the Cohen strategy
of deflecting cooperatives from protest into a collaborative role
was successful. Membership increased eightfold, and tonnage of
crops handled rose from 14,300 to 89,308, valued at 19,000,000
pounds.[9] The cooperative unions, normally organized on a district basis, acquired considerable importance. There were 14
ginneries in the hands of cooperative unions by 1962, numerous
staff was employed, and in many districts outside of Buganda the

cooperative union was the most conspicuous institution in the region.[10]   In Bugisu, most politicians active either locally or nationally were associated with the Bugisu Cooperative Union. Matthias Ngobi, Minister of Agriculture and Cooperatives until his arrest in 1966, made his reputation as Manager of the Busoga Growers Co-operatives Union from 1958 to 1962, a period during which the Union was remarkably successful.   Felix Onama, Minister of Defense under Obote, attained visibility as the General Manager of the West Nile Co-operative Union from 1960 to 1962; during his tenure, the Union already had a ginning monopoly in the area, and achieved the exceptionally high (and profitable) throughput of 23,103 bales in 1960-61.   George Magezi, who was a Minister from 1963 to 1966, was President of the most consistently effective cooperative Union, the Bunyoro Growers Co-operative Union, from 1958 on.

In this phase, many of the cooperative union managers were former officers of the Government Co-operative Department; this had been true of both Onama and Ngobi, who then found the more spectacular stage of the cooperative a valuable launching pad. The encouragement now accorded by the government included many opportunities for special overseas training courses, useful both for the experiences and also for the prestige conveyed.

On the whole, the second phase was one of relatively warm relations between the government and the cooperatives.   The Co-operative Department itself expanded considerably, with many of its expatriate officers bringing to their labors a personal commitment to the ideology of cooperation.   The sense that government now to some degree supported their objectives, as well as the Co-operative Department background of some key cooperative personnel, facilitated the mood of comity.   The major exception was Bugisu, where a large deficit in 1957 led the government to intervene, imposing a European supervising manager until 1961.

The situation of the cooperatives changed in phase three, from 1962 to 1966.   Their autonomy seemed to be further enlarged by a new cooperative ordinance in 1963, which made more onerous the procedures necessary for government intervention.   The extremely rapid expansion of cooperatives during this period, particularly in the cotton sector, provided an enormous flow of resources within it.   The number of members doubled, and the amount of cotton handled tripled.   The total turnover in the cooperatives by 1965 was 30 percent higher than the total revenue of all the local administrations in the country, including Buganda.[11]   With the speed of growth of cooperatives, and personnel turnover associated with Africanization in the Co-operative Department, the capacity of the government to closely monitor the movement was diminished.   So also, for a time, was its inclination, during this period when participative politics were at their zenith and central functionaries were likely to cause trouble for themselves by interfering in local affairs.   Further,

aspects of the 1952 and 1963 cooperative ordinances were used by
cooperative officials under fire to shield themselves from government tutelage.

In many areas, a web of nepotism, corruption, and mismanagement developed during this period; indeed, the benefit in higher
prices which should have accrued to farmers from the progressive
displacement of the immigrant monopolies in cotton ginning became
the ransom of cooperative inefficiency. By 1964, the Busoga
Growers Co-operative Union, formerly headed by Minister of Agriculture Ngobi, had reached such a state of decomposition that a
committee of inquiry had to be appointed--although its very
harsh report was not published until after the jailing of the
Minister in March 1968, for purported involvement in an anti-
Obote group. The Busoga Union, prior to independence, had been
one of the most effective and highly praised. After 1962, the
management had lost control to the elected committees. Effective
use of committee office involved use of influence in seeking employment opportunities for one's clientele; this process is well
described by the Committee of Inquiry, which noted, "a shame-
lessly high degree of nepotism and corruption in the appointment
of staff to positions in the Union; including cases of individual
committee members approaching individual managers and instructing
them to give employment to their relatives." The effects of the
committee intervention are detailed:

> the practice has grown up in the Union whereby junior
> staff, by virtue of some special relationship with mem-
> bers of the committee, can secure the sympathetic ear of
> committee members and report on their superiors. The
> "in-laws" network can be quite intricate in a polygamous
> society, and as Busoga can, by no means, be said to be a
> monogamous society, the chances are that each supervisor
> has under him some person who is to some degree an "in-
> law" of a committee member. It is for this reason that
> the managers are often afraid of some of their subor-
> dinates. The position is aggravated by the fact that,
> as the managers do not appoint their own subordinates,
> they cannot dismiss them; and because of their own inse-
> curity of office, have hardly the courage to discipline
> them.[12]

A rising tide of rural discontent over the operation of cooperatives in a number of areas struck home to President Obote,
when by 1966 a flood of complaints were heard from his own district of Lango concerning nonpayment for cotton delivered to
primaries. A Committee of Inquiry was hastily dispatched, and
found that, indeed, the practice of offering farmers chits rather
than cash on delivery of cotton had become widespread; far more
serious was the fact that in many cases the receipts were then

never honored. The Commission concluded that a large number of growers in fact had not been paid, and that this dereliction "was the result of sheer dishonesty and inefficiency on the part of the officials of the Primary Societies" who were "absolutely incompetent to handle large sums of money."[13] These findings were particularly unsettling, as the Lango Cooperative Union—reputed to be one of the most effective—had achieved remarkable expansion since 1960, with a turnover of 1 million pounds and an accumulated surplus of 1 million shillings.

A broader investigation into the whole cooperative movement produced an extensive catalog of inefficiency and corruption, and a series of recommendations for the escalation of government control and supervision. By the end of 1966, three of the largest unions (Busoga, Bugisu, Uganda Growers') were subjected to government intervention, and new cooperative legislation was on the horizon. The free-wheeling era of cooperative autonomy had come to a close.

In phase four, from 1967 on, control over the cooperatives was progressively tightened. The Busoga Committee of Inquiry had stated the matter bluntly: the government, under the 1952 and 1963 legislation, had found itself "as powerless as the members to institute any preventive measures against incompetence and malpractices . . . . At the same time it is obvious that the Government has invested vast sums of public money directly in the movement . . . . While respect must be paid to the principles of co-operation, a formula must be devised by which when the Government has more money invested than the members in a union, Government officers should be given overriding powers where major decisions are concerned."[14] A way, indeed, was found; in 1968, the Government announced that major revisions in cooperative legislation were planned, and in 1970 these were enacted. However, long before the formal adoption of the new legislation, extant powers began to be exercised far more vigorously.

An initial squeeze—reducing the volume of resources floating freely within the cooperative structure—was a sharp reduction in the allowances for transport and ginning costs in arriving at the price formula for purchase of lint cotton by the Lint Marketing Board. This began in 1965, in order to permit a rise in the price paid to the farmer. The move appeared dictated by the anticipation at that time that national elections might be held in 1966 or 1967, as required in the independence constitution. It was also stimulated by the discovery of the scale of excess profits in ginning in the terminal colonial decade, and a view that placing the cooperatives under greater pressure to operate efficiently might be quite salutary. Its effect was to accelerate the trend to financial crisis in the cooperatives, which in turn tended to spotlight the widespread malpractices within them.

The sharp increase in government supervision strengthened

the hands of the managerial staff, and greatly weakened the elected committees. The opportunity for determining recruitment policy, and constituting clientage networks within the staff became severely limited. Closer financial controls and more vigorous auditing by Co-operative Department staff made much more difficult the placement of intercept channels in the cash flow circuits of the cooperatives, through loans, tenders, expense accounts, and diverse other means. Thus the currency of committee membership was devalued during this period. The level of committee activity slackened; a somewhat dispirited apathy tended to replace the intense competition for committee places of the early independence years. Bunker summarizes the impact on the Bugisu cooperative Union:

> The BCU Committee's loss of powers has been accompanied by a steep decline in both leader and general membership involvement with the Union. Committee meetings, since the withdrawal of the supervising Manager in 1968, are held less frequently than they ever have been, and tend to be pro forma approvals of decisions made by the Secretary Manager and the Senior Cooperative Officer. . . . In contrast to the frequent meetings held, of visits made in the rural areas by all other Union committees, the present committee has never held a meeting other than in its Mbale offices, since it took office in 1966 . . . . The Union is evolving into an extension of the government rather than an autonomous agency.[15]

Such indeed was the dominant intention among the top political leadership. Government perspectives toward the cooperatives, and the general disappointment at their level of performance, were expressed with particularly brutal candor in 1967 by then Cooperative Minister Adoko Nekyon, a relative of Obote who at that time was an influential member of his immediate entourage:

> . . . democracy cannot work in a union. It can never work. If anybody is under illusion that democracy can be introduced into business, then he can take over from me even tomorrow morning . . . .
> . . . we are going to continue with dismissing of unions that are not functioning properly. Any Union which I find has been working badly should assume in advance that it is already dismissed.[16]

Government tutelage did stabilize the cooperative performance from an efficiency standpoint. Some of the largest unions, such as Uganda Growers, were split into smaller units, which appeared more manageable. More generous cost allowance formulas eased the pressure on cooperative ginneries and coffee factories,

and the scale of corruption appeared to subside. Overseas technical staff who had worked closely with the Uganda cooperatives
were of the unanimous view in 1972 that the level of cooperative
performance had been greatly improved.[17]

Only a very tentative appraisal of the evolution of cooperative affairs in the later Amin years is possible. The formal
machinery of cooperation was not dismantled; indeed, remarkably,
it continued to expand, with membership increasing from a half
million at the end of the Obote era to 1.1 million in 1978.[18]
Cooperatives operated some 54 ginneries and 51 coffee factories;
they continued to control a monopoly on cotton marketing. A significant private sector continued in coffee marketing, particularly in Buganda. In addition, a new underground private coffee
trading network (magendo) emerged to organize the smuggling of
an estimated 50-60,000 tons to Kenya (or about a third of total
production).

The rise in membership was accompanied by a sharp decline
in crop volume processed, especially for the cotton cooperatives
where farm output had plummeted. The prices paid to cooperative
unions for their ginned cotton, by the late 1970s, was less than
their ginning costs. Accordingly, they had sunk deeper in debt
to the banks (their bank overdrafts totaled 200 million shillings
by the end of 1978),[19] and farther in arrears in payments to
farmers. Their vehicle fleet, indispensable for buying operations, had dwindled, and spare parts for their processing plants
were often impossible to obtain.

At the same time, the complement of employees continued to
draw their salaries. One device used by cooperatives to sustain
themselves was to branch into diverse commercial activities which
apparently generated enough revenue to help meet the staff payrolls. The Commowealth team studying economic rehabilitation
after the Amin overthrow found cooperative unions operating
"ranches, pineapple plantations, various seedmills, bookshops,
hotels, rented properties, primary and commercial schools, consumer goods stores, engineering works, petrol stations, and dis-
pensaries."[20] One may surmise that the cooperative district
unions had become, to a large extent, mercantile cartels serving
their employees. Membership had little means of influence, and
the state supervision, which had grown strong in the preceding
phase, had atrophied in proportion to the demoralization of the
administrative apparatus.

COOPERATIVES, SOCIAL CONFLICT, AND MOBILITY

We may now return to the question of cooperatives as arenas
for social conflict and mobility. Here we must take note of the
three levels at which cooperative organization took place. The
basic unit was the primary society, where the direct interaction

between the movement and the farmers occurred.   In 1965, there
were 1,825 primary societies, with an average membership of 247;
by 1978, there were 3,054.   Their prime function was to buy and
bulk the cotton and coffee of the farmers; some additional functions might also be carried out, such as provision of credit and
some inputs, like insect sprays or fertilizer, or purchase of
other crops.   The primary employed a secretary-manager on a salaried basis, and perhaps a couple of other workers to help with
buying and storing the produce.   It was governed by a committee
elected by the membership, which had to meet at least annually.
The secretary-manager was likely to be either a primary school
teacher or a petty trader; sufficient literacy was required to
keep books.   An average cotton primary might have a cash flow of
50,000 shillings ($1 = 7 Uganda shillings), a substantial sum at
the village level.   Many primaries were coterminous with the low-
est echelon of organized administration, the parish.

   The primaries belonged in turn to regional unions.   In 1965,
there were 31 of these, and 41 in 1978.   The unions owned and
operated the processing units, ginning or coffee curing works,
plus warehouses, fleets of trucks and vehicles.   Their assets
might well total several hundred thousand pounds; a representative larger union might have a cash flow of 1 million pounds or
more.   The manager of a district union operated a substantial
enterprise; he was responsible to a committee elected by the
general assembly of primary society representatives.   The union
employed a number of permanent staff, and many dozen seasonal
employees to work in the ginneries and curing works for several
months following the harvest.   Union committee members were paid
for attendance at meetings, and received some travel and maintenance expenses; the committee chairman was a powerful figure, not
infrequently in conflict with the manager.   Through the 1960s,
it was quite frequent for unions to employ Asians as ginnery or
curing works managers.

   Finally, at the national level, there were three apex organizations: the Uganda Cooperative Alliance, intended to serve as
national spokesman for cooperative interests; the Uganda Cooperative Central Union, designed to serve a wholesaling role in acquiring supplies to be used or retailed through the unions; and,
from 1969, a cooperative bank.   In practice, these organisms have
had very little visibility or significance, for reasons to which
we will return.   Thus the essential components of the Uganda cooperative movement are the unions and primary societies.

   It was at the primary society level that the cooperative
interacted directly with the structure of village society.   At
an ideological level, it had frequently been argued that the cooperative could build upon the natural communitarian harmonies
and egalitarian thrust of village society.   In reality, the local
community was often riven with conflict and at least latent inequalities.   These cleavages were of several orders.   Often there

was a degree of rivalry between local lineage segments. In areas like Acholi or Kigezi, religion was likely to be a basis of discord, particularly the Protestant-Catholic division. By the 1960s, primary schools were widely distributed, and their teaching staff, as well as local traders, formed a local elite. There were invariably a small number of "Big Men," whose greater control over labor resources permitted them to farm somewhat larger tracts, and perhaps engage in some trade or transport as well. The micro-bureaucracy of local chiefs formed a different order of hierarchy, while elders generally might command more resources than the young. It would be absurb to deduce from this an endemic set of conflicts, but it is equally naïve to assume that there was a natural law of social harmony. The primary society in this setting, constituted a new and significant arena for social interchange, and was a funnel through which resources flowed. Stress rather than natural harmony was the normal condition of the primary society.

It may be reasonably argued that the recrudescence of rural violence and crime, commonly known as <u>kondoism</u>, was one social indicator of this strain. Kondoism began to provoke expressions of concern in the Legislative Council in the late 1950s, and continued to escalate in the 1960s. In 1971, President Amin invoked the kondo crisis as the justification--at first plausible--for the creation of the presidential Public Safety Unit, which then evolved into a terror strike force. Though kondoism had no direct relationship with the cooperative movement, its increasing prevalence caused grave difficulties in safely transferring the large amounts of cash from the district unions to the primary societies for payment of farmers.

In the highly politicized environment of the early 1960s, party labels often served as a vehicle through which local factionalism was expressed. The government energetically preached apoliticism in the cooperatives, to dampen this pattern of conflict. Despite the UPC aspiration for a one-party state, it never imagined that the cooperative movement could be captured as a party organ, as did the CPP in Ghana. A perusal of cooperative department district reports in these years makes clear the scale of the phenomenon; the lament of one West Acholi government cooperative official conveys the flavor of the times:

> It is no longer a hidden fact that whenever an election of a new Committee of a society takes place, politics is brought in, and as a result new Committees if at all elected properly are opposed by their political rivals. In the end, the Cooperative staff is faced with two opposing committees, each one declaring itself to be the true one for the society. It does not end up there; the Cooperative staff cannot tell which is legally right unless he attended the meeting where one was elected.

> Often the old Committee refuses to hand over the books
> and cash to the new Committee. There comes a deadlock
> which only a politician can unlock. Once the Cooperative
> staff tries to indicate which is the legal Committee ac-
> cording to the societies' by-laws, he is labelled as fa-
> vouring whichever party the legal Committee belongs to
> . . . . Appeals for political neutrality to members seem
> to mean nothing to them.[21]

This does not mean that primary societies were paralyzed,
and unable to operate. There is no evidence that significant
amounts of cotton or coffee failed to be marketed in areas where
a cooperative monopoly existed; in Buganda, there were many alternative private channels. In the final analysis, all parties
had a strong interest in assuring that the marketing function was
ultimately performed, whatever the frictional overheating of the
machinery in the process.

The primary society was thus an important arena for local-
level conflict. There was also a clear tendency for the societies to reflect extant structures of leadership. Joan Vincent
ably illustrates this process in a primary society in Teso, demonstrating at the same time the changes that occurred in the
third and fourth phases of cooperative development we have
sketched above. In phase three, the cooperative had been an important instrument for reinforcement of the position of the "Big
Men" of the parish, for most of whom cotton growing was only an
ancillary activity. "The cooperative society was, therefore,
less important as a means of improving agricultural production
than as a means of access to cash that could be invested in non-
agricultural enterprises. It was political office in the local
society that was valued--and jealously guarded by the Big Men
and elders--for its contacts with the power structure beyond the
parish, its potential political clout, and its kickbacks."[22]
In phase four, after the imposition of a cooperative monopoly in
cotton marketing in 1969, and acquisition of the local ginnery
in 1969, power shifted to a new set of officials who came in from
outside to operate the ginnery. They purged the staff and committee of the primary society, and supplanted the local "Big Men"
as the leaders--and prime beneficiaries. "No longer," Vincent
concludes, "is there merely the parochial exploitation of common-
ers by Big Men, themselves subject in the end to the leveling
mechanisms built into a rural system of competition. The exploitation is now that of a political class of urban-centered bureaucrats and politicians."[23]

The district unions were the most visible echelon of cooperative activity. This saliency was enhanced, in the second and
third phases of development, by the coincidence of cooperative
union and political district boundaries. The major cooperative
unions were headquartered in the district seats, often only a few

steps from the district government offices.   The districts in
these years were the most volatile focus of political life, and
largely served to shape contemporary ethnocultural identities.
By the early 1960s, the cooperative unions had a larger cash
flow, in most cases, than did the district governments.   In the
small, sleepy district seats, high cooperative office placed one
at the very top of the status ladder, along with district government leaders, and the small cadre of central officials attached
to this outpost.[24]   For cooperative leaders who made their
reputation in the late 1950s, there was the opportunity to cata-
pult to national prominence through the 1961 and 1962 parliamentary elections.   This pathway had already closed, however, with
independence, though few recognized it at the time.   Elections,
the indispensable institutional medium through which cooperative
leadership could be converted into national political standing,
were never again held until 1980.   From 1967, government encouraged the split of district unions into smaller bodies, significantly diminishing their visibility.

The most imporant single cooperative union, whose often
free-wheeling operations provided important redistributive opportunities, was the Bugisu Cooerative Union.   Committee membership
in this era was eagerly sought and bitterly contested; one former
committee member reported that election expenses could run as
high as 20,000 shillings.   A committee of inquiry complained of
"unbridled extravagance" reflected in "a fantastic increase in
. . . subsistence and travelling claims, which rose from 5500
shillings in 1960 to 62,827 in 1965."[25]

Stephen Bunker has produced fascinating data on the relationship between BCU office-holding and other patterns of mobility at this time.   While committee members, he found, were in
general established local politicians before securing election
to the committee, there was a clear tendency for them to enter
higher-ranking political or more highly capitalized business positions after committee service.   Four of the six who served as
BCU president achieved positions of substantial wealth and prestige within the district; none of these had more than six years
of formal education before they joined the committee, so this
mobility would not have been available in more structured careers
in the civil service where rank is explicitly tied to level of
education.   Some 46 of the 57 men who had served on the committee
since the establishment of the BCU in 1954 were involved with
significant business undertakings.   Of the three M.P.'s elected
in 1962, one was BCU Treasurer, a second was elected to the BCU
committee after winning his parliamentary seat, and the third
tried at that time to organize a rival coffee marketing associa-
tion--and twice ran unsuccessfully for the BCU chairmanship.   The
effective closure of political mobility channels in the years
after independence meant that the BCU spiralist had to link his
committee functioning with some kind of trade, as well as farm-

ing. For these purposes, the advantages conveyed by committee membership, in addition to the cash return, primarily related to the opportunity for frequent paid travel to Mbale, where commercial connections could be maintained and necessary transactions made.[26]

As the district unions became large enterprises, often operating a number of processing units, regional disputes frequently arose within them, with regard to siting of operations, or conflicting interests of those producing different crops. In Bugisu, coffee growers on the mountain claimed the cotton growers on the plain formed smaller primary societies in order to outvote them at the general assemblies; this was resolved by splitting the cotton operation off into a separate union. In Bukedi, heated arguments arose in 1959 over an inflated price paid for a ginnery, which opponents charged mortgaged the union to the benefit of those served by this antique facility.[27] In Busoga, the primary societies of the western side of the district argued that those in the east had resorted to the device of maintaining smaller primary units to obtain voting leverage in the union; this was converted into favoritism in staffing the union.[28]

The dynamics of regional conflict were well illustrated in the Bunyoro Growers Cooperative Union in the mid-1960s, at that time the best-managed Union in Uganda. With a turnover of 1,035,000 pounds in 1966, it was by far the largest economic enterprise in the district. At that time, five of the nine members of the Union executive committee were actively involved in Bunyoro district politics; the sectional and religious cleavages which characterized district council politics also flowed into the Union. The major issue was the proposed establishment of a tobacco factory. The two small towns of the district, Hoima and Masindi, which served as hubs for defining sectional factions, each laid claim to the new facility. Hoima, district seat, argued that 85 percent of Bunyoro tobacco was grown in its region. Masindi advanced the claim that a plant sited there could also process tobacco from neighboring Lango; the corporation which proposed to erect the plant in partnership with the cooperative union backed the Masindi site, maintaining that Bunyoro proper could not produce enough tobacco to ensure sufficient volume for the factory. Although the Masindi faction had a narrow majority, the rancor generated by the dispute so divided the Union that it could not secure finance, even from the government Uganda Commercial Bank. In the end, no factory was built, and the Union was seriously weakened by the struggle.[29]

Thus, in sum, for a brief period, the cooperative movement was an important political recruiting ground, and pathway for swift upward mobility. After independence, a period of slack government controls opened a range of opportunities within the political structures of the cooperatives—the elected committees —for a multi-faceted strategy of mobility, both through building client nets within the organizational structure, and through inter-relating resources obtained from cooperative activity with commercial activity, other employment, and some farming. The growing assertion of government control greatly reduced the significance of committees; power shifted to a bureaucratic coalition of Co-operative Department functionaries and managerial staff within the cooperatives. This also reduced the scope for political conflict within the cooperatives, especially at the union level where government tutelage was most directly felt.

COOPERATIVES AS POLITICAL ORGANIZATIONS

We may now turn to the question of the political impact of the cooperative movement. Cooperatives had become economic institutions of first importance; they were arenas for social conflict and for a time mobility. Had they also become important political vehicles in any meaningful sense at the national level? Had they become part of a nascent web of differentiated institutions binding the individual to society through an associational matrix of growing complexity?

As political organizations, cooperatives reached their peak very early in their development. In the interwar period, there was a steady stream of protest against cotton marketing policy in Buganda, and to a lesser extent Busoga; the proto-cooperatives in Buganda played an important part in this, and were able in 1934 to secure a promise from the Governor that attention would be given to formulating cooperative legislation. This, as we have seen, came to naught through the intervention of immigrant interest groups.

The zenith of cooperative activity in the political realm came through the Uganda African Farmers' Union (UAFU) in 1948-49, which was reconstituted as the Uganda Federation of African Farmers (UFAF) in 1951-52. Rural discontent was in crescendo over the low cotton price, government accumulation of huge reserve funds through holding down the producer price, Asian cheating in cotton purchasing, and the (correct) conviction that huge profits were being made in ginning. Officials believed that the exceedingly small cotton crop in 1947-48--a mere 166,000 bales, compared with the prewar peak of 418,000--was partly a consequence of this malaise.

The Uganda African Farmers' Union traces its own history to several antecedent organizations in 1947; however, these were smaller and less significant than the Uganda Growers Cooperative Union. The galvanizing of the UFAF into a potent organization came through a decision to seek redress to a range of grievances through political action; the invitation was made to the most visible political figure then available to take up the reins of leadership--Musazi. The Luganda press published an open invita-

tion to all peasants in Buganda in honor of Musazi, a man held
to be a vessel of God for the uplift of poor people.[30]   Musazi
himself, just returned from several years of deportation, was
anxious to launch a political movement.   His first effort, in
November 1947, had been in the less promising field of industrial
organization; he had inserted an advertisement in the papers
calling for all workers to organize themselves into local committees, in preparation for the foundation of a comprehensive national union.   Urban labor, however, was still a very peripheral
phenomenon; the real social energies lay in the countryside.

The Musazi scheme was to collect on credit peasant cotton,
then force government intervention to require that it be
ginned.   Preparations went forward to handle the 1948-49 crop on
this basis; although the government announced a price increase
from 22 to 30 cents per pound, this was denounced as still inadequate.   Gambuze, a Luganda paper, expressed a typical view:

> We the growers are aggrieved to see that the Protec-
> torate government refuses to free our cotton for which
> we sweat.   Probably the Government is so shy that it has
> been so far unwilling to inform us of the fact that we
> the producers are its labourers to whom it pays a wage
> or gift of Shs. 30/- on every 100 lbs. and reserves for
> itself Shs. 47/70 on every 100 lbs.   Once again we im-
> plore the Government to understand that it has no right
> whatsoever to save for itself any money that would ac-
> crue to us from our cotton.[31]

The Buganda chiefly hierarchy was mobilized by early 1949,
as buying season began, to warn peasants that those consigning
their cotton to the Musazi movement were likely to receive nothing in return.   In the end, most cotton did find its way into
regular channels, but not before a delay of nearly three months
in ending the buying season; the crop wound up more than twice
the size of the previous year.   The UAFU had made important headway in diffusing an awareness of what appeared to many Ganda
farmers as the iniquities of the cotton system.   Although the
disturbances in April 1949 fell far short of a real revolt, it
was the most sizable uprising the Protectorate had experienced
since the days of imposing its writ upon the territory.   While
the Government commission was not persuasive in documenting its
charge that the UAFU had provoked the disturbance deliberately,
there can be no doubt that the politicizing impact of the UAFU
campaign for peasant control of their own crop had created a pro-
pitious environment for the transformation of what began as a
large demonstration into widespread incidents of violence.

In 1951-52 the Uganda Federation of African Farmers enlarged
its scope of operations, and became more explicitly political.
The UAFU had been wholly confined to Buganda; now the UFAF estab-

lished small but politically significant beachheads in the Eastern Province, and Lango. Its net of grievance was cast over the
coffee sector as well as cotton; robusta, it will be recalled,
was spreading rapidly in Buganda, and was at this juncture a far
more remunerative crop. However, the small number of European
and Asian estate operators were permitted to process and market
their crop themselves, while Africans were forced to use immigrant processing facilities, and market through controlled channels. Further, until 1952, there was a racial pricing system,
whereby non-African coffee fetched a higher price than African
coffee. Although this was justified as a quality differential,
in practice, to Ganda coffee growers, it looked, felt, and
smelled like racial discrimination. The substantial export tax
was levied only on African coffee, on the argument that non-
Africans paid an income tax.

Some politically sympathetic expatriates, most notably
George Shepherd and John Stonehouse, were recruited by Labour
M.P. Fenner Brockway to assist with the management of the organization. By 1952, the UFAF claimed a membership of 80,000, although Stonehouse observed on his arrival that they had only
three old trucks, and a tumble-down office. Stonehouse,[32] far
more disabused by his experience than Shepherd, found the movement riddled with swindlers and a huge disappointment. "Promises," he wrote, "were the biggest stock in trade. The members
had an incredible faith in these promises although, so far, not
much of material value had been achieved. They believed their
leaders and thought that the millenium was just around the cor-
ner."[33] Shepherd, far more charitable, noted the indissoluble
link between cooperative and political aims in the eyes of its
leadership: "The Federation, to Ignatius's mind, was both a cooperative and a movement of the peasants aimed at one pre-eminent
goal--freedom for his people."[34]

With the arrival of Sir Andrew Cohen as Governor in early
1952, commissions were set up on cooperatives and the coffee and
cotton industry. Shepherd and Abu Mayanja, then an articulate
student leader, prepared a "memorandum on the Cotton Industry,"
which Shepherd described as the "first fully documented African
case." The cooperative commission visited a number of rural
centers, and took testimony from hundreds of farmers; the UFAF
assiduously stimulated testimony, and armed witnesses with specific objections to the existing cooperative ordinances. Although government regarded the organization as an irritating
nuisance, Musazi as a scheming opportunist, and Shepherd as an
unendurably naïve visionary, there can be no doubt that the reservoir of rural discontent which the UFAF tapped was taken seriously. The cotton commission called for government backing to
permit cooperative unions to acquire up to 20 ginneries. The
cooperative commission responded to many of the criticisms of
excessive restrictions in the 1946 cooperative ordinance, and

urged greatly enlarging the scope of autonomy for cooperatives, ushering in phase two as suggested earlier. The coffee commission followed by calling for abolition of racial duality in marketing, a common coffee price by grade irrespective of the race of the producer, and application of the export tax to all coffee.[35]   Prices for both cotton and coffee were permitted to rise to their world market levels, then very favorable. The UFAF leaders next diverted their energies into the Uganda National Congress; the farmer organization broke apart into component units, which separately registered with the Co-operative Department. With the high prices and prosperity of the mid-1950s, agrarian issues lost their sharp edge. Not long thereafter, in 1953, the political issue of the Kabaka's deportation eclipsed all else. Political parties became the pivotal institutions; never again would a cooperative be so politically significant as a protest vehicle acting upon national policies as was the UFAF.

In the 1950s and early 1960s, cooperative unions, then constructed primarily on a district basis, did offer an important framework for the ethnic self-awareness which became sharply politicized in the terminal colonial period. Although ethnic groups such as the Teso, Acholi, Gisu, and Soga had little if any sense of collective identity at the time of colonial penetration, the administrative as well as the church organization of Uganda followed an ethnic metaphor, and a sense of cultural identity did come to cohere about the district units.[36]   In the 1950s, the establishment of cooperative unions coterminous with these identity patterns made of this institution an embodiment, along with the local government, of ethnic community. This was most spectacularly visible in the cases of Bugisu and Sebei, which deserve some detailed consideration for the light they shed on cooperative structures as vehicles for ethnic self-affirmation.

In Bugisu

The Gisu were, historically, a series of localized communities, organized by lineage, with no centralized political institutions. The cultural potential for the subsequent emergence of a sense of Gisu unity was to be found in the similarity in dialect, the ecological commonality of Mount Elgon, similar customs, and a common set of circumcision rites, practiced bi-annually, which remain a crucial event in Gisu social life. A superstructure of chiefs suitable for colonial administration was imposed through Ganda intermediaries, and the administrative frame of a district (including the unrelated Sebei) provided the structural basis for crystallization of ethic awareness.[37]

Arabica coffee became the central feature of the Gisu economy during the colonial era. In the 1920s, resentment began to build against the Asian and European private traders who bought the crop, especially after the buyers formed a ring in 1927 for the purpose of "eliminating wasteful competition," which, literally translated, meant forcing down the price. The predatory nature of the buying ring was such that the government itself entered the crop purchase picture in 1930; in 1931, the Bugisu Coffee Scheme was established to supervise the marketing of Bugisu coffee, with the ultimate aim of promoting a cooperative.

Although the scheme was nominally a responsibility of the Bugisu Native Administration, in reality the central administration combined with the three main trading companies to share and organize the market, process the crop, and regulate the prices. The leading role was contracted to A. Baumann and Company, a Lever subsidiary. In 1933, the Bugisu Coffee Scheme was given exclusive buying rights; it was operated by a European manager hired by the District Commissioner, and managed by a board whose five members included two Gisu representatives.

Although the Bugisu Coffee Scheme was, for the Gisu, an improvement on the buying ring, the more lucrative positions within it continued through the war to be held by Asians, Ganda, and some Europeans. In 1938, after a series of deficits, the actual collection of coffee was contracted by the Scheme to an independent trading company, the Bugisu Coffee Marketing Company. An additional source of Gisu irritation was the appointment of an all-European governing board for the Company in 1940, with the old Bugisu Coffee Scheme board, and its two Gisu members, becoming purely advisory.

After World War II, Gisu complaints about the structure of the Bugisu coffee industry became more vocal. Two more Gisu were added to the Scheme advisory board in 1944 and 1949; they were able to serve as a channel for transmission of some grower complaints, and to argue for employment of more Gisu by the Company. However, the European Board of the Bugisu Coffee Marketing Company remained in full control, and Gisu conviction that they were being cheated by the Company became more deeply rooted.

When legal provision was made for cooperatives in 1946, Gisu quickly appreciated the possibility that a cooperative union could be the vehicle for eliminating the Company, and immigrant control over their coffee. A number of influential Gisu became active evangelists of the cooperative idea during the early postwar years. Perhaps a dozen played really influential roles, all the sons of chiefs or relatively wealthy men who had achieved early mobility as ancillaries to the Ganda chiefs. The best known of these, Samson Kitutu, had been an evangelist for the Protestant mission, and a leader of the Bugisu Welfare Association, which dated from the 1920s as a body to promote Gisu interests. Kitutu subsequently became first President of the BCU.[38]

The first two primary societies were begun in 1946, and by 1949, 24 were operating. In 1950, because of the growing animosity of Gisu to the Bugisu Coffee Marketing Company, its contract

was not renewed, and the Bugisu Coffee Scheme resumed direct collection of coffee. The missionary labors of the founding fathers
of the BCU now bore fruit: in 1954, the Bugisu Cooperative Union
was formed, and took over the operation of coffee marketing from
the Scheme. Elgon arabica was identified in the political consciousness of the Gisu as "theirs"; the BCU offered an institutional framework for achieving control over their primary source
of livelihood. It was, at once, an agency of economic nationalism, and a comprehensive framework through which Gisu solidarity
could be expressed. The Government Commission of Inquiry appointed in 1958 when the Union had encountered financial difficulties noted that, "The Union has become a symbol of Bugisu
unity. Any proposals to reorganize it are liable to be opposed

Table 4.2

PERCEPTION OF GOVERNMENT COOPERATIVE RELATIONSHIPS
(N = 480)

	All Uganda (%)	Bugisu (%)	Uganda, without Bugisu (%)
Coops are run by government	47.7	34.2	50.4
Coops are separate	43.1	64.6	38.9
Don't know	9.2	1.2	10.7

Table 4.3

INFLUENCE OF COOPERATIVES
(N = 465)

	All Uganda (%)	Bugisu (%)	Uganda, without Bugisu (%)
Cooperatives are able to influence government	40.0	75.9	32.6
Cooperatives are not able to influence	29.5	10.1	33.4
Don't know	30.5	13.9	33.9

on emotional grounds, and the authors of any such proposals may find themselves accused of 'divide and rule' motives."[39]

After independence, the BCU retained for a period great vitality as a spokesman for district coffee interests. Its President, Waisi, mounted an active campaign for the prerogative of marketing Busigu coffee directly on the Nairobi market, rather than through the agency of European or Asian trading companies, and to construct a 90,000 pound hotel and office block in Mbale. The saliency of the cooperative in the life of the district is well demonstrated in the results of a survey of farmer attitudes toward cooperatives and government services undertaken in eight districts, including Bugisu, by the authors in 1966.[40]

The sharp deviation of Bugisu from the all-Uganda patterns

Table 4.4

PARTICIPATION IN COOPERATIVE ACTIVITIES
(N = 256)

	All Uganda (%)	Busigu (%)	Uganda, without Bugisu (%)
How often do you attend cooperative meetings?			
Always	31.2	46.8	24.6
Usually	26.9	40.3	21.2
Sometimes	34.0	10.4	44.1
Never	7.8	2.6	10.1

Table 4.5

ABILITY TO IDENTIFY NAME OF THEIR DISTRICT COOPERATIVE UNIONS
(N = 232)

	All Uganda (%)	Busigu (%)	Uganda, without Bugisu (%)
Able to name	65.0	94.7	51.8
Unable to name	35.0	5.3	48.2

Table 4.6

ABILITY TO IDENTIFY LEADERS OF DISTRICT COOPERATIVE UNION
(N = 245)

	All Uganda (%)	Bugisu (%)	Uganda, without Bugisu (%)
Do you know the names of any of the officers of the District Cooperative Union? Who are the officers you have heard of?			
Names more than one	44.9	83.8	28.1
Names one	13.9	12.2	14.6
Unable to name any	41.2	4.1	57.3

in responses to questions relating to member perceptions of the
efficacy of cooperatives is striking. Nearly two-thirds of Gisu
respondents viewed cooperatives as organizations quite distinct
from government; in view of the bitter conflict between the Union
and the government over direct BCU coffee marketing and the Mbale
hotel, this is hardly surprising. Further, Gisu had a very optimistic opinion of the ability of cooperatives to influence government; less than a third believed this to be true elsewhere,
whereas three-quarters of the Gisu respondents thought it to be
so. Participation rates at that time were extremely high; 87.1
percent attended meetings most of the time, and nearly half always did so. Equally remarkable was the degree of familiarity
of the populace with the Union and some of its main officers.
It must be recalled that actual membership and participation was
at the level of the local primary society; the Union was one step
removed from the direct experience of the farmers. Yet its activities were well known, its communications network quite intensive, and its leaders household names. No less than 96 percent could name at least one BCU officer, and 83.8 percent knew
more than one name. Most organizations in the industrial world
would be gratified by such figures.
    The exceptional potency of the BCU was also given paradoxi-
cal recognition by the brief DP regime in 1961-62, which encouraged the formation of a rival grouping, the Bugisu Coffee Marketing Association. The BCMA was led by the mercurial S.G. Muduku,
who doubled as a member of parliament, elected in 1961 as UPC
and 1962 on the DP list. Muduku toured the district, promising
a 50 shilling bonus per pound of coffee; the BCU was forced to
counter with a 40 shilling bonus, which appeared to verify the
Muduku charge that the union had been underpaying the farmers.
The DP government granted the BCMA a buying license; Muduku trav-

eled to Europe and North America, and claimed to have negotiated advantageous sales contracts. Bitterness ran high, manifest in widespread violence and tree-slashing. The BCU and UPC charged that the BCMA was really a front for Asian interests; one of the first pieces of legislation adopted under the UPC government was the abrogation of the BCMA buying license, restoring the BCU monopoly. The BCMA was probably heading for bankruptcy in any case; Muduku showed his political agility by crossing the carpet to the UPC.

In Sebei

Not only political party, but also ethno-regional tensions divided the cooperative movement in Bugisu. The militant demands by the culturally distinct Sebei to secede from Bugisu district were paralleled by insistence on a separate cooperative union. Throughout the colonial period, Sebei had been a mere county in Bugisu District, on the north side of the mountain. Like the Gisu, the Sebei grew arabica coffee, which they marketed through the BCU. The Sebei ethnic population numbered only 33,000 in the 1959 census, or only 10 percent of the Gisu total. Sebei are of a wholly different language family than the Gisu, and are very conscious of their distinctiveness. However, the status of Sebei became an issue only in the late 1950s, as impending decolonization and political competition sharpened ethnic sensitivities. Indeed, the arguments which began to be advanced by Sebei leaders were the precise echo of those used by the Gisu in their successful demand for control of their coffee industry. Sebei complained that they were forced to market their coffee through the BCU, yet had no real voice in it; that almost no Sebei were employed by the BCU; that Gisu gave priority to their own interests, and short shrift to the Sebei. Separate district status was accorded in 1962. Separatism was then extended to the economic realm as well: Y.K. Chemonges, leading district political figure, lobbied hard in Kampala for authorization to secede from the BCU, which he finally obtained in 1964.

The Sebei District Council then voted to require all primary societies in the new district to belong to the Sebei Elgon Cooperative Union. One primary refused to join; situated on the Bugisu border, its membership was mainly Gisu settlers who had migrated into Sebei in recent decades. Chemonges, in a 1964 independence day speech, declared that Gisu farmers unwilling to join the SECU should leave the district. Out of this dispute grew the coffee war between Bugisu and Sebei, with widespread riots, many cases of arson and slashing of crops, several deaths, and the flight of several hundred Gisu out of Sebei. By 1967, the hour of euphoria was over, as the costs of cooperative self-determination were proving very high. The Sebei Union had to process its coffee at the Bugisu mill, and quickly found itself in very pre-

carious financial straits. By 1967, the SECU had a 1,000,000
shilling deficit; its volume was small, and management costs
high. In that year, returns on some 25 percent of coffee deliveries were never paid to the producers. The other side of this
coin was a fall in volume handled from 800 to 400 tons, much of
the balance presumably representing illicit sales to the BCU
where probabilities of payment were higher.

In Buganda

     In passing, we may note that the one place where coopera-
tives were most conspicuously not closely bound to ethnic sub-
nationalism was in Buganda. Both within the cooperative movement
and outside of it, a plurality of movements existed; only the
old UAFU and UFAF came close to covering all of Buganda. In the
Masaka area, where coffee is far more important than cotton, two
substantial unions devoted to coffee processing emerged, one of
them with distinct Catholic overtones. In the 1950s, a substantial private African sector existed in buying and processing.
At the time that provision was made for cooperative unions to
acquire ginneries in the cotton sector, in the robusta coffee
field the opening was made for private undertaking as well. Of
the first six African curing works established, four represented
private associations of growers, usually centered around a prominent and wealthy member of the Ganda elite. The associations
of growers became an arena for symbiosis with the non-African
private sector. In some instances, banks required that Asian
managers be hired in order that credit be extended. Many believed that, in a number of cases, these associations had become
conduits for Asian enterprises. Whatever the veracity of these
allegations, the existence of a wealthy landowning Ganda elite,
added to the limited influence of the central government in Buganda affairs (especially from 1962 to 1966), meant that a substantial private sector competed with the cooperatives. Thus,
although cultural sub-nationalism was more developed and intense
among the Ganda than with any other group in Uganda, it was
linked with the political institutions of the Kingdom, and not
the cooperative movement.[41]

At the National Level

     However, at the national level, what stands out is the very
slight political influence that cooperatives have had, despite
the scope of their economic operations. Some major national
figures passed through the cooperatives on their way to national
prominence. But cooperatives, as associational forms, have simply not spoken with authority as presumed spokesmen for their
constituency. The policy which most directly affects the membership is price; here the government has been guided primarily by

calculations on the world market, and the desire to protect its revenue position. On the one occasion since independence when a cotton price was set far beyond what the world market forecasts would suggest was prudent (and led to a 4,600,000 pound trading loss for the Lint Marketing Board in 1965-66), the governing factor was the prospect of national elections rather than pressure from the cooperatives. Without blinking an eye, it was quite possible--at a time when parliament was still an active force, and opposition parties were vocal, even if dwindling--to decree a brutal cutback in transport allowances and the ginning formula which greatly affected cooperative profitability. At the same time, quite high levels of export duty were maintained, amounting in 1967 to 25 percent of the export value of coffee and 14 percent for cotton. A commission of civil servants in 1966 investigating the cotton industry coolly recommended that the price to growers be reduced by one-third, a recommendation which was at once accepted. Sixteen cooperative unions submitted memoranda or testimony to the committee, but on crucial points there is no visible sign that their views were decisive.[42]

One answer to this apparent puzzle lies in the shifting interactions between three definable groupings connected with the cooperative movement: the cooperative politicians, the cooperative managers, and government cooperative specialists. In the early phases of the proto-cooperative movement, the influence of the cooperative politicians, who blended economic organization with political protest, was in the ascendant. In the 1950s, after the government removed many of the most blatant grievances with respect to price, rigid immigrant domination of buying, racial discrimination in robusta marketing, and the restrictiveness of cooperative legislation, the movement was, to a large degree, coopted. In key unions, members carefully trained for years on the Co-operative Department staff were placed in key managerial posts. In the Bunyoro Growers Cooperative Union, the key leader was A.N.W. Kamese, who served as a government agricultural officer in Bunyoro from 1932 to 1947, had a year of training at Loughborough Cooperative College in Britain, then served an additional tour as government cooperative officer before entering the Union full time. M.M. Ngobi, highly effective Manager of the Busoga Growers Cooperative Union in 1958-62, when it generated huge surpluses, had also worked with the government Co-operative Department from 1948 to 1957. Felix Onama was a government cooperative officer for ten years, from 1950 to 1960, before taking over as General Manager of the new West Nile Co-operative Union. C.M. Wakiro, the key managerial person in the Bugisu Co-operative Union from 1956 to 1966 (and later a fatality of the Amin regime), had also served his apprenticeship in the Co-operative Department, and had done a tour at Loughborough.[43]

The government cooperative specialists, until independence, were expatriate at the top ranks; there was a self-selection process in their recruitment, and those who entered this career stream--which was far from the most prestigious branch of government service--often chose it out of genuine commitment to the idea and ideology of cooperatives. Misty-eyed recollection of the Rochdale pioneers perhaps came less frequently after some years of daily combat with the everlasting dragon of corruption. All would have nodded in agreement wth the Eklund dictum quoted earlier, that the cooperative movement could be built only from the government outwards, while conceding that at some undefined future point they would have to assume greater responsibility for themselves. This set of promises was largely internalized by their Ugandan successors once independence had come. The first years of independence, and momentary diminution in the capabilities of the government cooperative specialists, gave an opportunity for the cooperative politicians, in the form of the elected officers and committee members, to gain in many instances a dominant position with respect to the cooperative bureaucracy.

Illustrative of this process was the sequence of events in Busoga. Ngobi had received close support from a senior government cooperative officer, Neal, who did much of the accounting work for Busoga Union, and in effect was seconded to it. Though Neal departed shortly before independence, another senior government officer, J.G. Miles, continued to work closely with the union. When he retired in 1962, an experienced Ugandan civil servant took over as government cooperative officer in Busoga. In the laconic words of an official report, this official, Kibenge, "was murdered during the hours of darkness" in early 1963. A series of successors were unable to maintain the tutelary influence of Neal, Miles, and Kibenge,[44] until the state moved in the late 1960s to sharply circumscribe cooperative autonomy. The managers, alarmed at the erosion of their power, then entered a coalition with the government cooperative specialists (from whose ranks many had come) to demand greater statutory government control over the cooperatives. Because managers and government functionaries were operating within the same normative framework, it was much easier for the managers to accept supervision from the Co-operative Department than what they viewed as persistent interference by the elected committees. The 1966 commission illustrates this point in its summary of testimony placed before it:

> We have received contradictory evidence with regard to whether the Registrar's powers should be increased or curtailed. Some unions, particularly their members of staff, suggested that the Registrar's powers should be increased to ensure effective control of the Co-operative Movement. Other unions, particularly Board Members, loathed the granting of extra powers to the Registrar. The reasons motivating such contradictory proposals are,

of course, obvious. On the one hand union staff feel
insecure in their appointments and would like the Registrar to protect them from ruthless boards. On the other
hand board members want to entrench further their powers
and build a kingdom of vested interests of their own.[45]

The colonial government vigorously propagated during the
years following the demise of the Musazi UFAF the postulate that
cooperatives should be apolitical. This was readily accepted by
the cooperative managers, and indeed was not really contested by
the cooperative politicians, nor by the dominant political party
after independence. Such father figures of the cooperative movement as Samson Kitutu in Bugisu were strongly hostile to political parties.
    It is interesting to note that in Bugisu, with the most po-
litical of all cooperatives in terms of the vitality of popular
participation within it, and the intimacy of the linkages of
leading BCU figures with the politics of the Bugisu District
Council, conflict within the BCU did not correlate with party
cleavages (which in any case were mutually fluid). The BCU was
vaguely associated in the public mind with the UPC, yet S.K.
Mutenyo, who replaced Kitutu as President in 1958 and held office
until 1962, became President of the Democratic Party in Bugisu.
At the same time, he had been a close ally of the best-known Gisu
politician on the national scene, J.N.K. Wakholi, a UPC member
of parliament, in a crisis over tax assessment in the District
Council in 1960. Mutenyo was succeeded as BCU President by
X.M.M. Gunigina, also a DP politico, in 1962. Ironically, the
BCU was led by Presidents associated with the DP at the time the
national DP government awarded a license to the BCMA. In elections for the BCU committee, candidates never used party labels. In fact, party politics in much of the country were simply
too divisive even for the ruling party to really pursue systematically the politicization of the movement.[46] The strong
overlap between the DP/UPC cleavage and religious identity as
Catholic or Protestant/Muslim would have meant that invocation
of party appeals would have at once politicized religious diver-
gence at the local level.[47]
    Officially, a national role for the cooperatives was to be
assured by two apex organizations--the Uganda Cooperative Alliance and the Uganda Cooperative Central Union. The former was
designed to serve as a national voice, and to promote cooperative
education. The Alliance was staffed almost single-handedly by a
well-known Ganda figure, Aloysius Kintu, beginning in 1962. Its
activity by way of offering a national cooperative viewpoint was
imperceptible. It did offer some 92 one-day courses to 9,240
persons in 1964, but these fell to 14 the following year. Many
cooperative unions, which were supposed to finance the Alliance
through membership and dues, resented what they took to be Ganda

hegemony in the organization, and either never joined, or declined to pay their dues. Kintu was also the operator of the
White Nile nightclub, a favorite Kampala watering spot during
the Obote years for the political elite; the combination of those
obligations with Alliance duties was a heavy burden. By 1966,
the Alliance was moribund. The Uganda Cooperative Central Union,
at that juncture, was almost equally paralyzed. Its functions
were supposed to be central purchasing of farm supplies and implements. The top leadership, in this case, was drawn from Bugisu; the Buganda unions were particularly critical of the UCCU,
making allegations of inactivity and ineffectiveness comparable
to those heard elsewhere about the Alliance. The government intervened in both cases to offer greater administrative support
and to require closer supervision, and both bodies improved their
record as service organizations at the end of the 1960s. By 1970,
the UCCU had a turnover of 20,000,000 shillings, and handled ginnery supplies, office equipment, chemicals and fertilizers.[48]
In our survey, some 32.8 percent of the cooperative members had
heard that some nationwide cooperative body existed, but only
6.2 percent could name one. As both became virtually state agencies, whatever potential they might have had as a voice for farmers disappeared.

    Whatever limited impact cooperatives may have had in the
Obote years as channels of rural representation, it is clear that
this ceased altogether under the Amin regime. The depoliticization
of cooperatives was well nigh complete. There was, of course, a
close relationship between the cooperative managers and the growing cadre of government cooperative officers, as long as the cooperative service remained immune from the slowly spreading decay
of the state bureaucracy. However, neither of these groups had
real access to the military. The general atmosphere of arbitrary
caprice dictated a risk-averting strategy of avoidance of conspicuous initiative which might draw attention to oneself.

    One of the structural channels defined for cooperative voice
was through representation on the marketing boards, especially
the Lint Marketing Board. The 1966 Cotton Commission devoted
some interesting observations to this issue. There was a widespread feeling at that time that the cooperative representatives
on the board were not particularly effective, and were more concerned with internal staff issues and redistributive matters internal to the Board's resource flow than with issues of general
policy--in short, that the governing board behaved in a fashion
analogous to the cooperative society committees. The Committee
of Inquiry, composed of civil servants, concluded:

        It was agreed that the reason for having a predomi-
        nance of growers on the Board was largely one of public
        relations--it was important that growers be informed of
        how "their" cotton was being marketed.

> The Committee considers that the primary function of such a board is not to act as a group of public relations officers . . . . Six people who have some contact with the cotton farmers are not likely to be very effective. We consider that members of the Board should be appointed according to their qualifications, experience, knowledge and ability in commerce and/or business administration.[49]

At that juncture, the government was not prepared to endorse quite such a technocratic view, and this proposal was rejected.

The role of cooperatives, as a new rural institutional framework, was thus paradoxical. Where well established, there is no doubt that they were significant local agenies. The confirmation of marketing monopolies in the key cash crops enhanced their importance; cash farming necessarily brings the smallholder into contact with the cooperative--although, so far, membership is not obligatory, and far from universal. The domestication of cooperatives by the government, the triumph of the manager functionary coalition made the movement more a vehicle for government "penetration" than farmer autonomy. From our present perspective, reading of trends suggests that cooperatives have had only limited success as institutional expressions of micro-community solidarity and integration. The liberal premises of the political development models bear the implicit assumption that the rise of new institutions implies differentiation. For the moment, the stake of the government in their efficient functioning appears to preclude this assumption being validated in practice.

5. Uganda Cooperatives as Government Policy Instruments

We turn now to consider the value of the cooperative structure as a policy instrument of the government, in enlarging its
institutional capacity in dealing with the rural sector. We will
explore this dimension by a review of two of the most highly promoted rural policy innovations of the Obote years--rural credit
and group farms. Both were designed to make use of the cooperative infrastructure, and thus constitute good tests of the role
of cooperation as a vehicle for state rural policy initiatives.

Rural credit and group farms both derived from a common conviction that a major breakthrough in farm productivity levels was
indispensable. For this to occur, the use of new inputs and mechanical implements deemed necessary. Access to credit was a
prerequisite for mechanization; group farming provided a convenient organizational device for incorporating tractors into a
peasant economy. Handling of a large volume of small loans un-
secured by land title could be done only through cooperatives.
The primary society also seemed a cogenial format for the group
farm. We examine these experiments in turn.

RURAL CREDIT

Historically, cotton and coffee production had developed almost wholly without institutional credit. The private banking
system confined its lending virtually exclusively to European and
Asian enterprises. Outside of Buganda, only a negligible amount
of land was registered, although in its last years the colonial
administration made some desultory efforts to encourage individ-

ual land titles. Thus farmers had no collateral to secure loans, and in any event the banks were not in the least interested in dealing with large numbers of small borrowers. In Buganda, where somewhat over half of the land had been incorporated in the mailo property system, and a significant fraction of the remainder had been more recently granted in individual tenure, lack of land title was not an impediment. In fact, the small volume of rural lending to Africans before 1962 was in practice limited to Buganda. However, even here banks discovered that it was very difficult to resell mailo land seized through foreclosure.[1] Some rural lending occurred through friend and kin, and small shopkeepers extended some credit. However, there was no class of moneylenders, and rural debt was not a significant problem.

While this array of bare facts would not have suggested the urgency of a government credit scheme, the project becomes more understandable in the context of the mood of impatience and hope which attended independence. Promises of loans and tractors played no small role in campaign rhetoric in these years, and rural audiences were very receptive to what were understood as pledges of tangible material benefits. Further, the effort had never been made; who could prove that it might not provide a new dynamic for rural development? Those in the public service who were skeptical, particularly the expatriate officers, and the Makerere agriculture faculty, found it prudent to mute their criticisms.

The first attempt to provide institutional credit to African producers was the Progressive Farmers Scheme, initiated in 1961.[2] Aimed, as the name suggests, at particularly successful farmers rather than the mass, the program was administered by the Department of Agriculture in conjunction with the Uganda Credit and Savings Bank. Repayment problems began to plague the program soon after its initiation.

In parallel to the Progressive Farmers Scheme, the Uganda government began to investigate alternative approaches. In 1960, it invited J.C. Ryan, rural credit specialist attached to the Reserve Bank of India, to inquire into the feasibility of a major scheme of landing to smallholders. He reported favorably and proposed a plan by which most credit would be in the form of direct supply of implements and other inputs, funneled to the farmers via the primary cooperative societies. The tie to cooperatives was intended to ensure recovery of loans through the deduction of repayment from the proceeds of the sale of borrowers' produce to their cooperatives.

A pilot scheme, based upon Ryan's suggestions, was launched in 1961-62, and default rates proved very low.[3] Meanwhile, the situation of the Progressive Farmers loan scheme continued to decline. By 1964, the Progressive Farmers program had been suspended,[4] and the government was ready to embark on a major rural credit scheme, administered through the network of cooper-

ative unions and societies. Together with the related group farm
scheme, cooperative credit was slated to absorb one-fourth of
the resources available for agricultural development during the
Second Five-Year Plan period from 1966 to 1971.[5]
    The mood of optimistic commitment to the program is re-
flected in the Second Five-Year Plan:

> For several years . . . the Uganda Commercial Bank has
> been lending money to cooperative loan societies for re-
> lending to individual farmers. This scheme has been a
> great success. In 1962 when the scheme started there
> were only a hundred participating societies; now there
> are over three times as many. Also, the rates of default
> have been so low as to be almost negligeable [sic].
>   The provision of credit is essential if productivity
> is to be increased, as in most cases such an increase in
> productivity will require the use of new inputs that will
> have to be purchased . . . .[6]

    However, with expansion the scheme quickly turned sour.
Default rates began to soar, and it proved difficult to see any
impact on productivity. Vincent shows in her study of a Teso
primary society that the loans were concentrated on the more affluent farmers, who also had the best opportunity to evade repayment. Gestures toward vigorous collection campaigns might be
made in response to pressure from above; in one such sweep, in
1967, ten bicycles were confiscated from young farmers, but no
"Big Man" was likely to be successfully prosecuted.[7]
    A careful study by Diana Hunt in Masaka and Lango districts
laid bare a number of the problems. The availability of credit
was not closely tied to new inputs or services which assured the
farmer a higher return. In Masaka, the small tractors that were
promoted through credit proved of dubious profitability: actual
use was only one-third of the official calculations on which
profitability assumptions were based. By 1966 all the tractors
which had been acquired with credit by individual farmers were
out of service; the costs and logistical difficulties of maintenance had been vastly underestimated. In Lango, where a major
use of credit was for tractor hire services, Matthew Okai demonstrated that labor costs for the two plowing operations which
could be performed by tractor were only 60 shillings per acre
with the use of hand labor only, as opposed to 85 for the tractor
(a rental rate which recovered less than half the real operating
costs).[8]
    From experience in Uganda and elsewhere, it is clear that
once the default rate reaches a certain threshold, it skyrockets.
As farmers learn that dwindling numbers are making repayment, the
loan comes to be viewed as a grant. When the anticipated revenue
increases fail to materialize, repayment becomes very difficult

and any sense of moral obligation tends to evaporate. Further, with the swift pace of expansion, government staff found it impossible to provide any supporting supervision; Hunt estimates that only about 50 percent of the loans actually were used for productive purposes in the first place.[9]

There were two spheres in which the picture was different: credit for peasant cultivators of Toro and Ankole, and for flue-cured tobacco in West Nile. In both cases, particular circumstances obtained. The financial return on these crops were substantially higher than for coffee or cotton. The net return on tea was 1,000-1,500 shillings per acre, with a capital cost of nearly 2,000 shillings during the five-year period before full

Table 5.1

EXPANSION OF THE COOPERATIVE CREDIT SCHEME

Year	Number of Societies	Total Loans (shillings) (7 sh = $1)	Number of Borrowers	Number of Borrowers in Default	Total in Default (shillings)
1962/3	101	830,917	6,859	168	10,640
1963/4	250	2,438,366	15,143	441	41,548
1964/5	281	3,701,965	23,382	1,635	195,291
1965/6	358	6,097,003	33,827	4,702	691,391
1966/7	286	4,915,553	27,842	n.a.	n.a.

SOURCE: Diana Hunt, Credit for Agricultural Development: A Case Study of Uganda (Nairobi: East African Publishing House, 1975), p. 96.

output is reached. Cotton had a net yield of only 200-250 shillings per acre, and robusta coffee 500-800. The tobacco return per acre was comparable to tea; loans here went to groups of four farmers for the mutual construction of a curing barn. The value of the crop, and focused zones within which it was produced, made possible the provision of an extensive infrastructure of supervision.[10]

We may discern a turning point in cooperative credit in 1966, parallelling the turning point in the general tone of relations between the UPC government and the cooperative movement as a whole, which can be identified at roughly the same point in time. (See Table 5.2.) The years from the start of the program

Table 5.2

COOPERATIVE CREDIT SCHEME LOANS AUTHORIZED AND UTILIZED
(excluding Group Farm loans)

Year	Bank Loans Approved[a] (shillings)	Loans Utilized[b] (shillings)
1962/3	896,450	655,140
1963/4	2,668,750	2,223,299
1964/5	4,109,840	3,307,019
1965/6	6,204,450	5,357,721
1966/7	5,904,845	5,064,333
1967/8	7,178,560	3,788,495
1968/9	9,305,050*	6,253,473*
1969/70	9,974,372	n.a.
1970/1	11,158,042	n.a.
1971/2	19,908,197	14,964,045

SOURCE: [a] 1962/3-1968/9: Department of Co-operative Development, Annual Report for the Year ended 31st December,
1968, Mimeo. (Kampala: Department of Co-operative Development, 1969), between pp. 10 and 11.
        1969/70-1971/2: Department of Co-operative Develop-
ment, Annual Report . . . 1971, Mimeo. (Kampala: Department of Co-operative Development, 1972), between pp. 15
and 16.

      [b] 1962/3-1968/9: as above.
        1971/2: Department of Co-operative Development, An-
nual Report . . . 1971, p. 15.

   NB: Figures from the 1968 Report refer only to bank loans authorized for society use. Those from the 1971 Report apparently
include certain sums approved for lending by societies from their
own funds. During the 1962/3-1968/9 period, such lending from
own funds added a total of Shs. 3.6 million to the sum available
from bank sources.

   * 1968/9 figures not complete as loans were still being issued
to robusta coffee growers in Ankole and Masaka when the sums were
figured.

to 1966 were the honeymoon period of the credit program, as in government-cooperative relations in general. The loans authorized and, more important, the loans actually utilized, increased at a terrific pace. In 1966, the cooling of enthusiasm, as well as a consciousness of the difficulties encountered, began to find expression in the stabilization of the sums approved and actually disbursed within the framework of the program. In percentage

Table 5.3

COOPERATIVE CREDIT SCHEME

Year	Loans Approved (a)	Societies Approved (b)	Societies Making Loans* (c)
1962/3	98	98	104
1963/4	248	246	239
1964/5	304	294	269
1965/6	376	343	316
1966/7	333	308	249
1967/8	303	267	228
1968/9	312*	265*	256*
1969/70	290	249	n.a.
1970/1	403	353	n.a.
1971/2	500	478	n.a.

SOURCE: 1962/3-1968/9   (a,b,c): Department   of   Co-operative Development, Annual Report for the Year ended 31st December 1968, Mimeo. (Kampala: Department of Co-operative Development, 1969), between pp. 10 and 11.

1969/60 (a,b): Department of Co-operative Development, Annual Report . . . 1969, Mimeo. (Kampala: Department of Co-operative Development, 1970), p. 16.

1970/1, 1971/2 (a):   D.S. Frederickson, Cooperative Specialist-Credit, "November 1971 Monthly Report," Ms., Kampala, American Cooperative Development International headquarters files, Washington, D.C.

1970/1, 1971/2 (b): Department of Co-operative Development, Annual Report . . . 1971, Mimeo. (Kampala: Department of Co-operative Development, 1972), p. 15.

* 1968/9 figures not complete as loans were still being issued to robusta coffee growers in Ankole and Masaka when the sums were figured.

Table 5.4

COOPERATIVE CREDIT SCHEME: SUMS IN DEFAULT
(shillings)

Date	Total	Regular Credit Scheme	Group Farms	Tobacco
9/68[a]		1,081,900	890,141	
5/71[b]	3,112,000	875,000	900,000	1,299,659
10/71[c]	2,782,000	600,000	906,000	1,238,000
1/72[d]	4,463,000	600,000	900,000	2,900,000

SOURCE: [a] Department of Co-operative Development, Annual Report for the Year ended 31st December, 1968, Mimeo. (Kampala: Department of Co-operative Development, 1969), between pp. 10 and 11.
[b] D.C. Frederickson, Cooperative Specialist-Credit, "July 1971 Monthly Report," Ms., Kampala, American Cooperative Development International headquarters files, Washington, D.C.
[c] D.C. Frederickson, Cooperative Specialist-Credit, "November 1971 Monthly Report," Ms., Kampala, Agriculural Cooperative Development International headquarters files, Washington, D.C.
[d] D.C. Frederickson, Cooperative Specialist-Credit, "March 1972 Quarterly Report," Ms., Agricultural Cooperative Development International headquarters files, Washington, D.C.

terms, the growth of loans granted during the remainder of the Obote years was moderate. A parallel pattern may be noted in the figures on group farm loans granted.

In general, we may say that so long as the government was in need of backing against still serious KY and DP opposition, the support of cooperative leaders was highly valued, and resources were channeled to the cooperatives and through them to rural areas. With the forcible repression of Ganda opposition and reduction of the DP to impotence on the one hand, and the decline of the economic and technical functioning of processing, credit, and cooperative organization in general on the other, government felt both the need and the ability to act in a harsher, more punitive fashion. There was a sharp slowdown in the rate of program expansion. Indeed, the number of societies for which loans were authorized was steadily cut back. Credit

was cut off for those societies in default. (Compare 1965/6--
316 societies making loans valued at 5.4 million shillings,
versus 1968/9--256 societies making loans valued at 6.3 million
shillings.) By these means the professional staff of the Co-operative Department and their American advisors would seem to have
overcome the tendency for loan default to become a tidal wave,
and to have limited the spread of the attitude that loans could
be viewed as grants. It became clear that participation in the
credit program was a benefit available on a long-term basis, but
only to those maintaining reasonable repayment records. Short-
term gains from avoidance of repayment could only be bought
dearly, by sacrificing the longer-term benefits of continuing
participation in the credit scheme. The results in terms of
liquidation of back debts in the latter years of the UPC period
are summarized in Table 5.4.

The major exception to the introduction of a more sober,
businesslike approach to the administration of the credit scheme
during the post-1966 period was the unrestrained expansion of
tobacco loans in the northern districts of Acholi and Lango.
Even after the political changes of 1966, the UPC still felt it
necessary to provide special rewards to the heartland of party
backing. Once it became clear that the group farms (to which we
turn next) were an expensive fiasco, a more economically sound
alternative was sought. The success of closely supervised, cooperatively organized and financed tobacco growing schemes seemed
a possible solution. Thus, credit scheme eligibility rules were
waived for tobacco growing societies in Lango, Acholi, and other
northern areas, and large sums were streamed to farmers by this
means. Predictably, default rates were very high, and by the end
of the UPC regime, the accumulated group farm debts, together
with the deficits of the unqualified tobacco societies, were the
major default burden weighing upon the cooperative credit scheme
as a whole. By 1971 most of the default from the regular cooperative credit scheme framework had been liquidated, thanks to the
vigorous loan recovery efforts of the departmental staff and
their advisors.[11]

In 1971, in the wake of the overthrow of the UPC, there
once again a marked leap in loans granted. This seemed to reflect renewed confidence in the cooperative credit scheme and
the new regime's desire to provide tangible benefits to the countryside, much as the UPC had done in its earlier, more insecure
years. (Amin devoted much of his first year in office to barn-
storming tours of the country and meetings with representatives
of different ethnic groups and socioeconomic sectors of the population; in July 1971, he summoned together representatives of
farmers from all parts of Uganda to a large conference termed the
"Farmers Forum."[12]) Unfortunately, the documents available to
us do not permit extension of the repayment series into the Amin
years. In any case, the general shocks which destroyed Uganda's

economy in these years undoubtedly had their effect on repayment rates, an effect which has little to do with the use of the cooperative as the distributive mechanism for rural credit programs.

How should the cooperative credit scheme be evaluated in terms of the developmental criteria of equality and efficiency? Studies available suggest that though typical small farmers were able to participate and receive cooperative loans, a disproportionately large portion of loan funds were distributed to the better off and more powerful. Conversely, repayment drives were felt most strongly by the less influential farmers, while the affluent had the best opportunity to evade repayment.

With regard to efficiency, we have already noted that persistence brought an improvement in the overall results achieved in the repayment of cooperative loans. Thus, by January 1972, only some 3.5 million shillings were in default out of approximately 45 million shillings which had been loaned during the period 1962/3-1970/1. Through the transfer of responsibility for loan administration to the cooperative societies, government freed itself from the administrative and economic burden inherent in the processing of thousands of small loan applications and made possible a program which might well have been simply impossible were the full responsibility to fall directly upon government ministries and associated public financial institutions.[13] While noting that the cooperative mechanism made possible the loan program's expansion and that a reasonable recovery rate was achieved, we must point out that available evidence suggests that the credit distributed contributed little to the improvement of production methods and their technological advance.[14] Thus, what might be seen from the developmental point of view as the ultimate economic objective of cooperative credit programs was not achieved.[15]

However, we would suggest with regard to this point, that the failure to contribute to the transformation of agricultural production techniques had little to do with the choice of the cooperative as the administrative device for the distribution of credit. The potential of the cooperative as a framework for the finance of both agricultural innovation and accompanying agricultural extension activities existed. If the potential was not exploited the fault was not inherent in the cooperative mechanism itself, but rather must be seen as lying with the government departments (Agriculture and Cooperatives) responsible for the credit scheme. The absence of a portfolio of profitable innovations and the failure to coordinate credit and extension activities must be attributed to these departments, rather than to the cooperatives. As for the cooperatives themselves, under the supervision of the Co-operative Department they achieved the ability to fulfill their major, direct responsibility within the credit scheme--distribution and recovery of loans--in a reasonable fashion.

Remarkably, the Commonwealth team in 1979 found that the credit scheme was still active, and pronounced a quite optimistic verdict upon it:

> The scheme has, over the years, worked well. Through
> it farmers have been able to pay for essential inputs
> and associated with these there has been the opportunity
> for timely agricultural extension. Default rates have
> been relatively low, 4-9% in the case of farmers and 1.5%
> in the case of primary societies. The main problem with
> the scheme has been the inadequacy of the funds avail-
> able. . . . The total in the fund now stands at 15.8
> m[illion shillings]. As a result only 505 out of more
> than 3,000 societies are able to participate in the
> scheme.[16]

GROUP FARMS

Group farming was also a product of the mood of urgency and experimentation of the independence era. There had been an earlier surge of enthusiasm for group farming in colonial milieux at another moment of new departures, immediately following World War II. A number of such schemes were launched, particularly in Kenya, in the late 1940s; they swiftly failed, and official interest evaporated as quickly as it had emerged.[17] In 1962, the idea reappeared in a number of countries, partly inspired by the impressive success of analogous cooperative farms in Israel, in particular the Moshav ovdim.

The concept of the group farm was to bring together groups of approximately 100 farmers to exploit cooperatively a substantial block of land, averaging 2,000 acres. The government supplied 4-6 tractors per group farm, and made available supervisory personnel, including in a number of instances a European manager. The participating farmers were to form a primary society, in which was vested the title to the land, and the capital equipment. The members, however, tilled individual plots, and retained ownership of their produce, which was to be marketed through the primary society. Government paid for the initial clearing of the land, and financed the acquisition of the tractors and other equipment. Access roads and water supplies were also underwritten by the state. In the words of one government official, "These Group Farms are destined to become nuclei of integrated rural development schemes where Government and the members of the schemes will jointly undertake the tasks of land development and increased agricultural production; leading to the improvement of every aspect of rural life." The group farms were designed "to replace the existing pattern of small, unplanned peasant holdings."[18]

Expansive words indeed, but they do well convey the new
vistas of rural transformation which group farming seemed to
open. The first two group farms were established in 1963; by
1965, 31 had been launched. By this time, a number of problems
had already become visible; these were, however, ascribed to the
inevitable frictional problems in any new undertaking. Two expatriate government officers wrote in 1965 that early setbacks
"were largely due to insufficient planning."[19]   The authors of
the Second Five-Year Plan concluded that:

> The group farms . . . were initially faced with sev-
> eral difficulties. However, what stands out is not so
> much the difficulties but both the rapidity with which
> many of these have been overcome and the record yields
> achieved . . . . Group farms are, therefore, a most im-
> portant means of carrying out the necessary structural
> changes in the agricultural sector and, as such, the
> creation of new group farms is one of the most important
> projects of this plan.
>
> It is not just their economic advantages which justify
> so large an expenditure. They provide a concentrated
> nucleus of people united in one economic activity and
> thus make the provision of many services such as health,
> education, electricity and water more economic.
>
> In this Second plan, at least a hundred new farms will
> be set up in addition to the present forty . . . .[20]

Nor were these empty words: group farms were accorded the largest
single capital budget line in the Second Plan, some 4.29 million
pounds of a total 19.64 million pounds earmarked for agriculture.
    The group farm idea initially enjoyed not only enthusiastic
backing from the UPC leadership, but also from local politicos
and many farmers. Our survey showed that 59.6 percent had heard
of group farms; of those who had a clear opinion, 66 percent were
favorable. Numerous demands for group farms appear in parliamentary question time in the early 1960s; our interviews at the district level, and examination of district council and agriculture
records also confirm the extent of the interest aroused by the
project. As Colin Leys notes, by 1965 the policy had acquired
real momentum, although when it first emerged in 1963 it showed
signs of hasty concoction. Local politicians found that winning
a group farm for their area was an important patronage weapon;
of the first 19 group farms established, the siting of half was
determined by politicians. No less than 80 applications were
received for the 1965 quota of 20 new farms.[21]
    The group farm commitment was first viewed with some dis-
taste by civil service specialists, and especially the Makerere
agriculture faculty. However, by the time the Second Plan was
in preparation, group farms were a steamroller; few dared openly
oppose them, or do more than raise oblique objections on practi-

cal difficulties in implementation. Senior Ugandan public servants themselves became committed to their success, and government determination to pursue the experiment continued with inertial force for some time after the danger signals cumulated to
an unmistakable reading of disaster.

Inadequate planning was only one of a number of difficulties
which crowded in upon the scheme. The tracts of land required
were considerable, and problems were often encountered in iden-
tifying blocs of the requisite scope where adequate fertility,
water supply, and access to transport were assured. The ideal
of the cooperative farming community was rarely attained; the
bulk of the members on most farms did not take up permanent resi-
dence on the group farm, but maintained their plots elsewhere as
well. The attraction of the farms was above all the free clearing and highly subsidized plowing; this could well facilitate
worthwhile returns for a small commitment in the first season or
two. Many also discovered that if they grew food crops rather
than cotton on their group farm plots, they could benefit from
plowing services but escape deductions for loan repayment by marketing their produce through private channels instead of the cooperative. Often group farm members were not primarily farmers,
but teachers or traders for whom the cooperative plot was a side-
line. Only infrequently were the primary societies formed to
sponsor the group farm harmonious entities; conflict and cleavage
was frequent within the societies, as well as between the members
and the European farm managers.

Not only did the group farms fail to function according to
the idealized model, but their costs proved a considerable burden. Very few ever met their operating costs; most had poor repayment records on loans. The second plan target of 100 new
farms implied concentrating 20 percent of the agricultural investment on, at most, 10,000 persons, or less than 1 percent of
the nation's farmers. As their failure became widely apparent,
and a number were deserted by their members, it was clear that,
far from serving as nuclei of integrated nodes of rural development, their only demonstration effect was negative. There were
other hidden costs; the district agricultural officers were required to attend group farm meetings. A number of those interviewed reported a high levy from their available time diverted
into the group farm problems. They also began to be a political
liability; by 1965 DP parliamentarians were developing an attack
on this issue. Among their allegations was the charge that
senior officials received commissions on the large British tractor sales associated with the group farm project.

In 1967, a policy review of the whole sector of government
tractor purchase and group farming made manifest the multiple
deficiencies. The 876 tractors purchased, of which nearly one-
third had been assigned to group farms, had been very underuti-
lized, averaging only about 600 hours of service per year.[22]
Tractors were rented at a rate which required a subsidy of 13.63

shillings per hour in the year of maximum efficiency (1963); this implied a subsidy of 1 million shillings in 1967.[23]   At the same time, according to the Minister of Agriculture Kakonge, the policy review evidence "made it imperative to take a careful and cautious attitude in the establishment of more Group Farms."   The total peaked at 43 in 1966, and by 1969 had dwindled to 18, none of which were operating.[24]   Kakonge tried to put a positive gloss on the situation, by concluding that "there is an all-out effort to re-organize the old ones . . . as well as to encourage the formation of well planned new ones."[25]

The discontinuity created by the Amin coup provided the occasion for quiet burial of the group farm project, now through metamorphosis a case exhibit of the incompetence of the Obote regime.   A review of "achievements" during the first year of the second republic concluded that "The Group Farm Project was ill-conceived and doomed to failure."[26]   The third plan made no mention of group farms, while indicating a disillusionment with the whole thrust of mechanization, beset with a series of crippling difficulties.   The variety of makes made spare part supply and maintenance difficult; a number were not suitable to Ugandan conditions; control over use of a dispersed fleet meant "frequent misuse" and "inordinate . . . down time"; there was a critical shortge of mechanics, and the low rates charged to farmers meant large deficits.   "To make the high cost even more burdensome," the plan concluded, "there is little conclusive evidence that tractor use alone had generally increased the output of any agricultural product."[27]   However, the wheel came full circle in 1976, when President Amin announced that, as part of a "new departure" in agricultural policy, 600 new tractors were being purchased by the government.[28]

In contrast to the credit scheme, the group farm venture was an unmitigated failure.   The mechanization aim was ill-conceived from the start; its practical result was the squandering of substantial resources with little practical benefit to the society as a whole, in increased output, or even--except for some fleeting advantages for a season or two--for the small number of cooperative group farmers who were the target of this large public investment.   The Ugandan experiment demonstrated that the cooperative structure was not easily converted into collective production units.   Farmers were initially attracted to an illusory hope of high individual windfall gains from what they understood to be virtually cost-free, state-supplied mechanical tilling.   When these expectations were shattered, the units fell apart; member commitment to the concept of group farming was nil.   Rural cooperative production units have generally met the same fate elsewhere in Africa (Zambia, Mali, Guinea, Tanzania, for example).   The pressure for such experiments has never come from the farmers; the concept invariably originated in ideological projects fostered by the national elite.

COOPERATIVES AND AGRICULTURAL POLICY: RETROSPECT AND PROSPECT

In conclusion, we may note that, by the time the Amin regime seized power, a much more subdued view of the potential policy role of the cooperatives had become dominant within senior ranks of the government. Partly this was in reaction to the association of cooperatives with the Obote regime. In the third plan, in the key effort to diversify agricultural development beyond cotton and coffee, "the limited operational efficiency of the cooperative movement" was noted.[29] Its brief and ineffectual monopoly in food products was ended, and the coffee monopoly pledged by Obote was never fully implemented.

> The last regime is reputed to have attached great importance to the Co-operative Movement. Yet its important and fundamental principle of political neutrality was almost completely violated. Seeing that co-operative unions and societies were substantial, well organized bodies, some politicians quickly harnessed leaders in order to get the support of organised bodies for their selfish interests and genuoures [sic] for their supporters . . . . This resulted in splitting societies, and the formation of new and uneconomic groups based on political affiliations.
> The advice of the Co-operative Department was ignored by these politically appointed co-operative society officials, and many things went wrong resulting in the dissatisfaction of the co-operators and a lot of ill-feeling.
> The former regime failed to fight <u>kondos</u> who did havoc to co-operative funds, causing numerous farmers not to be paid cash on delivery of their produce, or not to be paid at all . . . . A number of cooperative officials who were known to have embezzled society funds received public support by the Government . . . .[30]

The survival of the cooperative movement in the face of pervasive institutional decay in the later Amin years is in itself a remarkable achievement. While the cooperative infrastructure has suffered the same dilapidation of its physical capital as have all sectors of the economy, it is nonetheless striking that there is unanimity within and without the country, among those who have addressed the issue of rehabilitation (or, more ambitiously, redesign) of the post-Amin Uganda economy that cooperatives will remain central to marketing and processing of at least the traditional export crops, coffee and cotton.[31] However far short cooperation falls from the visionary goals initially designed for it, the formula has become an institutional given, virtually taken for granted.

Cooperation, then, appears to be an integral part of the uncertain future facing this devastated country. It certainly op-
erates in an altered economic setting. Its most secure bastion,
the cotton trade, is threatened by the problematic future of this
crop, massively abandoned by farmers in the 1970s. Smallholders
have now become accustomed to producing mainly food crops, for
both consumption and sale. The inertial force which kept many
bound to the cotton economy has been severed. Whether the state
and cooperatives will be able to offer sufficiently attractive
prices and secure marketing arrangements to entice farmers back
to cotton is impossible to forecast. Both, however, have strong
institutional reasons to try: the state, though it suspended the
cotton export tax in 1977, still desperately needs the foreign
exchange from cotton sales and supply for domestic textile factories; the cotton cooperatives can survive only on adequate
processing volume for their ginneries. On the coffee side, the
loss of value of the Ugandan currency created tremendous incentives for smuggling; around this traffic emerged a whole new
parallel (magendo) marketing system, which now effectively com-
petes with the cooperatives. Magendo itself is so institutionalized, and has come to involve such an array of powerful post-
Amin personalities, that its elimination will be a slow and
difficult task.

Beyond the uncertainties of the political and economic environment of cooperation, what instruction for the future comes
from the lessons of the past? Certainly, the dilemmas will remain on the critical value issues of equality and efficiency.
These worthy objectives do not necessarily go together.

On the equality front, there is little to support the hopeful view that cooperative structures themselves operate in a redistributive pattern. The local primary societies tend to reflect community hierarchies. The district cooperative unions are
strongly influenced by the interests of their managers and employees, particularly since tighter government supervision meant
slackened participation. We have little basis for even speculat-
ing as to the inner dynamics of primary societies or cooperative
unions in the later Amin years; it seems implausible, however,
to imagine that earlier patterns were altered in this respect.

Nor can it be said that cooperatives have been very effective in realizing the distributive goal of transferring the middleman profits of the earlier private traders to the smallholder
producers. Cooperative operating costs, especially for the district unions, were too high, in good part because of the often
irresistible pressures they faced to expand their payrolls. More
disconcerting to the farmer was the element of unreliability in
the cooperative marketing system, which hit a peak in the middle
1960s. Many cooperatives made illegal deductions from payments
due farmers, for transport or other expenses. Most serious of
all was the practice of payment by chit, which often forced the

farmer to make repeated trips to the buying center in the hope
of having these receipts redeemed.   Cooperatives also suffered
substantial losses from embezzlement and theft.[32]

Faced with a perceived tradeoff between equality and efficiency in the middle 1960s, the state chose the latter, and
stepped up its supervision of the movement to impose it. We are
persuaded that significant headway was made; those most closely
connected with the cooperatives believe this to be the overall
trend. By 1969-70, the cooperative structures did prove capable
of purchasing, processing, and marketing record cotton crops,
and much of the coffee, with fewer complaints than in the middle
1960s. We should add that the effective state supervision oper-
ates mainly on the district union level; Co-operative Department
staff were never numerous enough to exercise close tutelage over
the 3,000 primary societies.

Perhaps the greatest cooperative success is simply survival.
At this juncture, we cannot imagine the deliberate dismantling
of the cotton and coffee cooperatives. Possibly, the primary
societies, as they become a prescriptive feature of local society, do become partially assimilated to indigenous values of reciprocity and exchange, and thus more valued institutions. More
sustained, patient government support can improve the efficiency
of the operation of the movement as a whole.

6. Uganda Cooperatives and Their Members: Farmer Attitudes

Cooperatives exist for the benefit of their members, at least ostensibly. The time has come to listen to farmer evaluations of the Uganda cooperative movement. These were expressed to us both through the mechanism of our farmer survey, through the qualitative assessment of our student assistants, and--indirectly--through our interviews with cooperative officers, government field staff, and other locally knowledgeable persons.[1]

From these sources, several clear themes emerge, which we will develop in this chapter. The powerful initial appeal for the cooperative alternative to immigrant Asian and European marketing channels, while well known, is further documented. Once established, cooperatives must demonstrate to their members--on a continuing basis--their economic utility; ideological appeals are not enough. The variations in farmer evaluation from one district to another are very great; these help illuminate the factors which contribute to cooperative success.

The 500 farmers interviewed were older than a normal age distribution in Uganda might suggest, although not so dramatically an aging population as cocoa farmers in Ghana. About half were over 40. Fewer than 20 percent had finished primary school. Two-thirds had farm holdings of less than 4 acres (interviewers made no effort to measure the size of the holding). Only 7 percent said they had been designated as "progressive farmers" by the Agriculture Department. Nearly two-thirds reported they had earned no money from nonfarm work during the previous year, though more than half had been employed off the farm at some point.

Table 6.1

AGE OF RESPONDENTS

	N	%
50 and over	129	26.2
40-49	109	22.3
30-39	133	27.0
29 or less	95	18.9
Don't know	26	5.3

Table 6.2

LEVEL OF EDUCATION

	N	%
Complete secondary or more	2	0.4
Complete junior secondary or partial secondary	20	4.1
Complete primary or partial junior secondary	72	14.7
Partial primary	190	38.7
None	206	42.0

Table 6.3

APPROXIMATE SIZE OF HOLDING*

	%
10 or more acres	13.6
5-9 acres	20.7
1-4 acres	58.5
Less than 1 acre	7.1

* Nearly one-fourth of those inter-viewed were unable to give a precise estimation.

Rural communities in Uganda are by no means isolated and encapsulated. Of our respondents, 20 percent had visited the capital of Kampala during the past year; 60 percent had been to their district seat, with as many as 30 percent reporting frequent visits (10 or more times). Another survey carried out at the same time as ours showed a surprising degree of media exposure. Of 421 respondents in 21 villages scattered through the country, half the villagers classified by the investigator as "rank and file" listened to the radio regularly, while 79 percent of the local influentials listened almost daily. Remarkably high levels of political information were discovered: some 97 percent of the influentials and 61 percent of the rank and file could identify Kwame Nkrumah; of these, most influentials and 44 percent of rank and file knew he had been overthrown a few months previous to the interview.[2]

Table 6.4

RADIO AND NEWSPAPER EXPOSURE
(n = 421)

	Radio		Newspapers	
	Influentials (%)	Others (%)	Influentials (%)	Others (%)
Never	3	13	11	31
Seldom	10	36	23	30
Once a week to 3 times a week	8	11	31	24
Everyday or almost everyday	79	39	32	13
No information		1	3	2

SOURCE: Anthony Oberschall, "Communications, Information and Aspirations in Rural Uganda," Journal of Asian and African Studies, 4, no. 1 (January 1969), pp. 30-50.

Our respondents included 46.0 percent Protestants, 36.3 percent Catholics, and 5.2 percent Muslims; the remainder primarily described themselves as "pagans." In relation to the 1959 census figures, showing Protestants and Catholics with almost equal numbers, Protestants were somewhat over-represented in our group. While there were no apparent systematic differences in attitude

which correlated with religious affiliation, there was evidence
of a culture of poverty associated with those not connected with
one of the major religions.   The "pagans" were invariably the
smallest and poorest farmers, and appeared to regard themselves
as a kind of lumpen-peasantry.   "I'm only a pagan," a farmer replied, in explaining why he would not expect a government officer
ever to visit his farm.[3]
    Slightly more than half the respondents (51.8 percent) were
members of cooperatives.   Of these, over half had been members
for five or more years.   Some 16 percent of the non-members had
once belonged to cooperatives.   Variation between districts was
high on membership levels: 76 of 78 interviewed in Bugisu belonged, whereas in Buganda only 14 of 50 respondents were members, though 17 others were former members.

Table 6.5

LENGTH OF MEMBERSHIP, COOPERATIVE MEMBERS

	N	%
Over 10 years	73	28.2
5-10 years	65	25.1
3-4 years	45	17.3
1-2 years	32	12.4
Less than a year	12	2.4
Former members	38	7.6
Don't know	44	8.8

MOTIVATIONS FOR MEMBERSHIP

    To understand the motivations for joining, we need to recall
first the peasant animosity to expatriate cotton and coffee buyers which took shape quite early; by the 1930s it was well recognized by the colonial administration.   "Cheating" was the specific focus of the grievance.   Around the well-nigh universal
conviction that Asian traders practiced systematic fraud, there
developed an institutionalized hostility, which was deeply rooted
by the time that cooperative promotion became official policy.
Against this setting may be understood both the general receptivity to cooperative establishment in the 1950s and 1960s, and the
enthusiastic reception of General Amin's brusque expulsions of

Table 6.6

ATTITUDES TOWARD ASIAN TRADERS

Private Asian traders carry out their business honestly:

	N	%
All of the time	16	3.4
Most of the time	7	1.5
Only some of the time	92	19.2
Never	352	74.0
Don't know	10	1.2

Table 6.7

ATTITUDES TOWARD AFRICAN TRADERS

Private African traders carry out their business honestly:

	N	%
All of the time	22	5.3
Most of the time	37	8.9
Only some of the time	125	30.0
Never	223	53.5

Asians in 1972. In our survey, 93.2 percent of the respondents felt that Asian traders carried out their business honestly only occasionally or never. At the same time, it should be noted that enthusiasm for private African traders was not much greater.

One may thus note, at the point of departure, a parallel with the early history of the cooperative movement in Europe. Although the antagonism to the private market mechanism in the Uganda case lacked ideological formulation, nonetheless one might argue that cooperation was an alternative to capitalism. The private trading system had served to institutionalize a racial specialization of function, with Africans confined to the least remunerative phase. The net impact of colonial regulation of the market had served to freeze this allocation of functions, and also guarantee the profitability of the expatriate sector. The

reservoir of discontent thus generated facilitated the swift expansion of cooperatives.

The Asians, however, were a wasting asset. However popular
any measure might be which removed them from the rural economy,
such a move could be made only once. Thereafter, the cooperatives had to develop some continuing claim to the loyalty of
their members. Fundamental in this respect is the capacity of
the organizations to deliver economic benefits--or at a minimum
not to impose an economic penalty on membership through inefficiency, nonpayment, or corruption.

In explaining their initial decision to join, those survey
respondents who were cooperative members suggested the para-
mountcy of expectations of economic benefits, as well as anti-
Asian sentiments. The reliability of the responses, of course,
is diluted by the retrospective nature of the explanations; they
are as much an indication of a present frame of mind as a description of past motivations. Nonetheless, the saliency of
economic factors stands out.

Aside from the Asian question, social and political objectives played little part in cooperative affiliation. As Table
6.8 indicates, fewer than 20 percent listed among their reasons
anything in this category. Early nationalism and regional affir-
mation played an important part in the early history of the movement, as we indicated in Chapter 4. However, by the time of the
full flowering of government-sponsored cooperation in the 1960s,
these motivations had receded far into the background.

Table 6.8

REASONS FOR JOINING COOPERATIVE

Why did you join the cooperative?

	N	%
Mentions bonus and other economic reason	32	12.4
Mentions bonus only	19	7.1
Mentions loans	2	0.8
Mentions other economic goods and services	81	31.0
Mentions elimination of Asian traders	71	27.0
Mentions promotion of African welfare	30	11.4
Mentions general social objectives	37	14.0
Mentions political or ideological objectives	14	5.3

(Percentages total more than 100 percent because multiple reasons
were recorded.)

POINTS OF FRICTION

The issue of the bonus is an interesting reflection of the psychology of cooperative appeal. The bonus was a second payment to members, distributed on the basis of profits from the marketing operation. Not only was the promise of a bonus frequently cited as a reason for joining, but failure to receive it was a significant factor in decisions to resign by former members. At the time of the survey, as indicated in Chapter 4, the cooperatives were placed in a profit squeeze by the government decision to maintain an artificially high price for cotton, achieved by a combination of sharp cutbacks in cost allowances made to ginneries for their transportation expense from local collection points to the ginnery, and in ginning costs themselves. Only very efficient cooperatives could maintain bonus payments, and many fell below this threshold. Only 22.3 percent (59) of survey respondents reported receiving a bonus during the past year; another 28.4 percent (75) had received a bonus in the past, but not during that year, while 48.9 percent declared that their cooperative had never paid them a bonus.

A related question was the payment of cash for crops on delivery. On close examination, one may appreciate that handling crop finance is a major organizational challenge, which many cooperatives could not easily manage. To pay cash means placing money in the hands of the farmers several processing stages before the cooperative receives payment for its ginned cotton or hulled coffee from the marketing boards. These involve cash advances, either from the government via the marketing boards--and no reserves remained to accomplish this--or from the banking system, which applied commercial criteria of managerial efficiency in granting advances. Thus the challenge was considerable, and it is hardly surprising that many failed to make the grade. Only 21.7 percent (57) of cooperative members reported that their cooperative always paid cash; 61.4 percent (169) stated they sometimes received cash, while some 14.1 percent (37) claimed that they had never been paid cash. It goes without saying that failure to provide immediate payment was a source of intense grievance--all the more since payment of the chits might be long delayed, or in a number of cases never made.

Corruption was another source of tension with the members. At the time of the survey, government inquiries established the depth and prevalence of dishonesty in the movement.[4] The very structure of the marketing process, which generally required substantial amounts of cash at a large number of buying points, made a degree of peculation difficult to control. Cooperative theory presumes that the democratic internal structure of the movement provides an adequate control mechanism; indignant members can vote dishonest committees out of office. In practice, this is often very difficult. Equally problematic was prosecution of

embezzlers: even though much of the actual labor of the Co-operative Department consisted of auditing the books of local societies, gathering evidence to make possible successful prosecution was a massive task; fraud at the society level was virtually
risk-free--comparable, say, to burglary in New York City. Further, much of the corruption lay in an ambiguous zone of impro-
priety, such as excessive expense allowances, dilation of the
payroll through nepotistic appointments, and manipulation of
credit.

In the face of other evidence as to the scope of corruption,
cooperative members had a surprisingly benign attitude toward
honesty in the movement, though less than a third would go so
far as to believe that their cooperative was always honest. The
evaluation was far more favorable than that of Asian or African
traders--though a substantial number sold part or all of their
crop to private dealers where cooperatives were failing to pay
the full price, or were not offering cash. On the other hand,
it fell far short of a ringing vote of confidence in the integ-
rity of the movement.

Table 6.9

ATTITUDES TOWARD HONESTY OF COOPERATIVE

Number feeling the cooperative is honest:

	N	%
All of the time	147	31.5
Most of the time	110	23.1
Only some of the time	157	33.0
Never	47	9.9
Don't know	15	3.1

Finally, the overlapping concept of "mismanagement" was another negative dimension of farmer relations with cooperatives.
This included issues such as overstaffing, which converted the
high profits of the private traders into redistributive channels
peopled by friends and relatives of committee members. Careless
management could lead to refusals by banks to advance money for
crop finance, forcing the cooperatives in turn to offer only IOUs
to members delivering their crops. The Committee of Inquiry into
the Affairs of the Busoga Growers Cooperative Union concluded
that the Union had lost some 1,500,000 shillings through buying

Table 6.10

EFFICIENCY OF COOPERATIVES

Cooperatives are not working efficiently.

	N	%
Agree	158	33.9
Disagree	221	47.9
Don't know	87	18.7

malpractices from 1962 to 1964. Perhaps the most spectacular example of inefficiency was the loss in 1962-63 of 800,000 shillings' worth of cotton simply through being left outdoors in the rain. In the words of the Committee of Inquiry: "The cause of this colossal loss was masterly inactivity and gross irresponsibility by everyone concerned with the Union from the Committee down to the ginnery manager. No employee had the common sense or initiative to take remedial action to prevent bales of cotton being damaged by rain for fear of incurring the displeasure of the Committee."[5]

Yet the overall assessment of the efficiency of cooperatives was, again, surprisingly positive; only a third of the entire set of respondents (including members and nonmembers) agreed with the statement that cooperatives were not working efficiently. However, the picture changes significantly if the two districts, Bugisu and Bunyoro, where confidence in the cooperatives was high, are removed from the picture. In these two districts, only 16.7 percent of the respondents believed that cooperatives were not working efficiently. For the remainder, the proportion of those believing the cooperatives inefficient rises to 43.5 percent.

This point is verified by the results of another survey undertaken in Teso in 1967 by Victor Uchendu and K.R.M. Anthony. They found that cooperatives "are plagued with poor management, accounting irregularities and sometimes deliberate cash theft or burglary of the cotton store." Despite this, "there was no lack of enthusiasm for the idea of the co-operative on the part of the farmers we interviewed."[6]

Other of our survey responses likewise indicated a moderate degree of satisfaction: some 51 percent felt the cooperatives had provided the benefits expected (Table 6.11). Only 20.4 percent felt private traders offered better prices than cooperatives (Table 6.16). Reasonable confidence was expressed in the caliber of the cooperative staff. As we note below, these replies become somewhat less impressive if broken down by district; if the very favorable evaluations in Bugisu and Bunyoro are removed, the remaining districts show a more jaundiced view.

Aside from a marketing channel for their cotton or coffee, a substantial number had received some other service through the cooperative. Some 35.9 percent of cooperative members had received a loan; 18.5 percent had obtained farm implements, and 33.8 percent had obtained sprays (insecticides). Nearly half had at least once obtained some husbandry information from the cooperative. While these figures are not overwhelming, they do suggest that the effort to utilize cooperatives as a multipurpose vehicle for rural service delivery had some positive impact.

REGIONAL VARIATION

Perhaps the most striking element in the survey responses was the wide variation among districts. Local circumstances clearly played a major role in shaping perceptions of cooperatives. Overall, Bugisu and Bunyoro stood out as having a high degree of member support and loyalty. In Buganda, former stronghold of the Uganda Federation of African Farmers and cradle of the cooperative movement, farmers had a much longer experience of one or another form of cooperation. Their rather negative assessment reflected the series of disappointments with various initiatives in the cooperative domain. In Kigezi, cooperatives were just being introduced; there was at once the favorable anticipation that the benefits promised for the movement might arise, and skepticism provoked by initial organizational difficulties. As Kigezi produced almost no cotton and not much coffee, it had been much less affected by the hostilities generated by the racial division of labor in agricultural marketing--and thus the appeal of eliminating Asians had less resonance. In Acholi, Busoga, and Teso, cooperatives were in the throes of severe internal crises when the survey was taken; echoes of these problems emerge clearly in the critical evaluations made by many respondents in these districts.

The Bugisu and Bunyoro cases stand out for the strikingly favorable evaluations of the cooperatives. In our fourth chapter, the relatively high belief in the efficacy of the cooperatives as a channel of influence, and the spectacular level of awareness of the district unions and their officers were cited. The following tables offer interesting insight into the contrasting perspectives of farmers in Bugisu and Bunyoro as compared with the rest of the districts surveyed. (In Tables 6.11 through 6.19, the questions were asked only of the 259 cooperative members. The N for each of these tables is approximately the same as that given for Table 6.11.) It is worth noting that there was substantial variation in the three Bugisu parishes sampled; in Central Bugisu, 25 of 26 responding to the question cited improvement, while in North Bugisu only two-thirds felt things were better, and in South Bugisu less than half.

A final set of evidence reflective of the striking contrast between Bugisu and Bunyoro and the others is found in the responses to a set of widely expressed criticisms of cooperatives. Although overall farmers tended to disagree with these standard complaints, the picture appears less favorable once Bugisu and Bunyoro are removed. In some districts, some of the complaints--especially about inefficiency and poor quality of cooperative personnel--attracted substantial backing.

Table 6.11

EVALUATION OF COOPERATIVE SERVICES

Has your cooperative given you
the things you expected when you joined?

	All Uganda (N = 248)	Bugisu (N = 80)	Bunyoro (N = 81)	Uganda less Bunyoro and Bugisu (N = 87)
Yes	51.0	90.8	72.6	25.2
Some of them	23.1	5.1	27.4	32.6
No	25.9	4.1	0.0	42.2

Table 6.12

PERCEPTION OF IMPROVEMENT OF COOPERATIVES

Do you believe that your cooperative is working
(better, the same, worse) than it was when you first joined?

	All Uganda (%)	Bugisu (%)	Bunyoro (%)	Uganda less Bunyoro and Bugisu (%)
Better	54.5	64.5	77.5	46.5
Same	19.3	25.0	22.5	16.1
Worse	26.3	10.5	0.0	37.4

Table 6.13

COOPERATIVE EFFICIENCY

Cooperatives are not working effectively.

	All Uganda (%)	Bugisu (%)	Bunyoro (%)	Uganda less Bunyoro and Bugisu (%)
Agree	33.9	15.4	17.7	43.2
Disagree	47.4	83.0	60.0	34.6
Don't know	18.7	1.6	22.3	22.2

Table 6.14

INSTABILITY OF LEADERSHIP

Cooperatives are always changing their officers,
and they never are able to learn their job properly.

	All Uganda (%)	Bugisu (%)	Bunyoro (%)	Uganda less Bunyoro and Bugisu (%)
Agree	26.7	15.4	25.9	29.8
Disagree	45.0	79.4	31.7	39.6
Don't know	18.7	5.2	42.4	30.6

Table 6.15

CONFIDENCE IN COOPERATIVE STAFF

Cooperatives cannot find good people to work for them.

	All Uganda (%)	Bugisu (%)	Bunyoro (%)	Uganda less Bunyoro and Bugisu (%)
Agree	34.0	20.2	20.9	41.2
Disagree	49.0	78.4	52.3	40.3
Don't know	17.0	1.4	26.8	18.5

Table 6.16

COOPERATIVES COMPARED TO PRIVATE TRADERS

Cooperatives are not giving as good prices as private buyers.

	All Uganda (%)	Bugisu (%)	Bunyoro (%)	Uganda less Bunyoro and Bugisu (%)
Agree	20.4	14.1	28.4	20.0
Disagree	74.4	80.9	70.5	73.8
Don't know	5.2	5.0	1.1	6.2

Table 6.17

GOVERNMENT INTERFERENCE IN COOPERATIVES

Government is interfering too much in the cooperatives.

	All Uganda (%)	Bugisu (%)	Bunyoro (%)	Uganda less Bunyoro and Bugisu (%)
Agree	21.4	36.2	7.7	21.4
Disagree	58.4	47.3	75.6	56.3
Don't know	20.3	16.5	26.7	21.3

Table 6.18

RELIGION AND COOPERATIVES

Some people are trying to drag religion into cooperatives.

	All Uganda (%)	Bugisu (%)	Bunyoro (%)	Uganda less Bunyoro and Bugisu (%)
Agree	18.9	11.5	30.9	17.5
Disagree	75.2	88.5	61.9	75.4
Don't know	5.9	0.0	7.2	7.1

Table 6.19

COOPERATIVES AND POLITICIANS

There are too many politicians in the cooperative movement.

	All Uganda (%)	Bugisu (%)	Bunyoro (%)	Uganda less Bunyoro and Bugisu (%)
Agree	20.9	12.9	22.6	29.8
Disagree	63.7	83.5	56.3	48.6
Don't know	15.5	3.6	21.1	21.6

Some distinctions between Bugisu and Bunyoro are worthy of note. Bunyoro did not differ markedly from all-Uganda figures on instability of officers, use of Asians as technicians, or presence of politicians in the cooperative movement. Indeed, Bunyoro respondents exceed the aggregate averages for criticisms concerning cooperative prices lower than those private traders offered, and the intrusion of religious conflict into cooperative affairs. Asian traders still played a significant role in Bunyoro by the time of the survey, whereas in Bugisu they had been completely displaced by the cooperatives. Religion, although less significant an element of division in Bunyoro than in Kigezi or Acholi, nonetheless was far more salient in local affairs than in Bugisu. The one complaint which found strong support in Bugisu was the allegation that government was interfering too much in cooperative affairs. At the time of the survey, the Bugisu Cooperative Union was locked in combat with the government over the right of the Union to market its coffee directly to Europe, and the construction of a hotel in Mbale. The BCU had been founded and had grown as an expression of local cultural sub-nationalism in antagonistic relationship to government, both before and after independence. In Bunyoro, where an exceptionally small number of respondents was troubled by the role of government, cooperative history had been quite different: the government had been a benevolent patron of the Bunyoro Growers Cooperative Union, and, as we have noted before, the father figure of Bunyoro Cooperation, A.M.W. Kamese, had served for many years as district agricultural officer. Thus there were some significant qualitative differences in the generally positive assessment of respondents in the two districts most receptive to cooperatives. In Bugisu, no less than 53 percent had been members of a cooperative for more than ten years; the cooperative movement was an institutionalized part of local life.
    Buganda was an interesting reflection of the problem facing

cooperatives after a prolonged series of failures and disappointments. No less than 34 percent of the respondents in the two
parishes sampled were former members, as opposed to 16 percent
nationwide. Cooperatives faced not only the competition of private Asian traders, but also an important sector of African entrepreneurs and large operators. Some 63.6 percent felt the cooperatives were never honest, compared to 31.4 percent nationwide. Only 27 percent were cooperative members; of these, only
one had voted in the local cooperative society elections. The
cynical view of cooperatives was part of a larger mosaic of suspicion of the national government; the belief that prices were
held down by government was particularly strong, and nearly all
felt the government did nothing for the farmer, compared to 38.9
percent nationwide.

Acholi, Busoga, and Teso also showed quite critical patterns
of response, illustrative of the difficulties then experienced
by cooperatives in these districts. It was a general practice
for cooperatives to levy "deductions" from the government price
for cotton, purportedly to cover transport or other operating
costs, though this was contrary to government policy. Religious
division cut deeply into cooperative politics; illustrative of
the kind of issues arising were charges by one religious community that cooperative societies dominated by committees of another persuasion refused to buy their cotton, or imposed other
disabilities.

The survey in Uganda occurred at a time when, except in
Buganda, moderate optimism concerning the future prevailed, which
must have vanished in the Amin era. This mellow mood was reflected by our respondents, who overall were relatively hopeful
about their future prospects, with nearly half expecting to make
more money in the years ahead. There were some hopes for better
prices for cotton and coffee, most planned to raise their output,
and many planned to select new crops. Evidence is found in the
responses that a smoothly running cooperative can make a significant contribution to a hopeful outlook on the future; there were
close correlations between optimism on the future, satisfaction
with the cooperatives, and receipt of a cooperative bonus.

Table 6.20

EXPECTATIONS FOR THE FUTURE
(N = 470)

	(%)
Expect to make more money in future	45.0
Do not expect to make more money	31.5
Don't know	23.6

Table 6.21

ANTICIPATED SOURCE OF HIGHER INCOME
(N = 212)

Which of the following reasons lead you to hope for more money
(asked of respondents who indicated higher income was expected).

	Yes (%)	No (%)	Don't Know (%)
Better prices for crops	52.5	44.3	3.2
Increasing production	87.1	12.4	0.5
Starting new crops	52.4	47.1	0.5

Table 6.22

COOPERATIVE SATISFACTION AND OPTIMISM

	Satisfied with cooperative		
	Yes	Some	No
Expect higher income, yes	91	32	32
No, don't know	28	22	28

(Correlation coefficient = .21; chi square signficant at .01.)

Table 6.23

COOPERATIVE BONUS AND OPTIMISM

	Received Cooperative Bonus		
Expect Higher Income	Past Year	In Past, but not Previous Year	Never
Yes	52	49	64
No, don't know	6	20	54

(Correlation coefficient = .30; chi square significant at .01).

CONCLUSION

For the Obote years, the picture that emerges from our data is of a systematic sequence of reactions to the growing fact of cooperative domination of agricultural marketing. After an initial phase of enthusiastic support, attitudes subsided into more passive toleration. Cooperatives that succeeded in providing efficient services to their members, or in active articulation of local views, were able to generate quite impressive levels of support. If performance reached a nadir where palpable costs were imposed on farmers--through nonpayment of cash, prolonged absence of bonuses, or sectarian conflict--a strong current of resentment could set in. However, if cooperatives could maintain a minimum threshold of effective economic performance, then farmers were prepared to accord at least a passive acceptance, if not active commitment. In the late 1960s, the main thrust of government policy was toward greater tutelage and control over the cooperatives, to ensure that these minimum standards were met. At the same time, the cost of this phase of more active intervention was the curtailment of the zone of autonomous cooperative activity which had provided the opportunity for the more vigorous unions, such as the Bugisu Cooperative Union, to earn a high degree of participation and commitment.

In short, what the farmers tell us is that the two crucial factors in their appraisal of cooperatives are the delivery of material benefits, and the reliable performance of marketing services. These findings, we may note, are fully consistent with studies of agricultural cooperation in the United States and elsewhere.[7] Thus the equality versus efficiency issue is perceived quite differently by the farmers than it is by the state elite or external observers. There is little evidence from our survey or interviews that egalitarianism within the local community, as an abstract question, comes into play. Cooperation in material terms is judged above all by the price paid and delivery of a bonus. The cutting edge of "efficiency," from the smallholder perspective, is being able to rely on cash payment at the stipulated price for the cotton or coffee when it is delivered to the buying point.

From the vantage point of the farmer, we can perhaps understand why material benefits and market reliability are much more important than flaws in cooperative operation more visible to the national policy-makers. For example, government cooperative bureaucrats seemed more exercised over the corruption issue than were farmers themselves. Bennett, in this connection, cites the phenomenon of "institutionalized suspicion" in the Western institutional model of cooperation that was transplanted in Uganda. The requirement of government audits of cooperative accounts is a reflection of this. "It is possible that this element," Bennett argues, "so typically Western in its synthesis of opposites, is the most exotic aspect of institutional cooperation from the point of view of agrarian people in other societies, where the system is self-policing due to social consensus, or by authority constraints built into the culture. . . . In fact, this element is usually viewed as dangerous since it sows distrust in a communal system."[8] This, as well as fear of reprisals, may help to explain the widespread reluctance of farmers to come forward as witnesses in corruption charges the Co-operative Department wished to see prosecuted.

7. Dairy Cooperatives in Uganda

INTRODUCTION

The milk marketing cooperatives established in Uganda from the mid-1960s on provide an interesting example of cooperative activity within a new branch of agricultural production, which first began to receive significant government encouragement in the post-independence period. As with cotton and coffee, the growth of cooperatives in the dairy sector was strongly influenced by the national political leadership's urgently felt need to intervene in the agricultural sector so as to gain politically essential control of economic resources and promote economic development. Moreover, there were economic-technical considerations which produced a degree of genuine farmer interest in the services of dairy cooperatives. Milk, like cotton and coffee, requires bulking for shipment to distant markets; and the cooling of milk, like the primary processing of export crops, requires an investment in equipment beyond the means of all but a tiny minority of estate-scale producers.

Thus, as we will show, dairy cooperatives have not been a purely bureaucratic imposition upon milk producers. Instead, in many areas an important segment of dairy farmers has seen cooperatives as important adjuncts to milk production and become significantly involved in cooperative activities. Our first task in this chapter, then, will be to examine the balance between material, solidary, and purposive factors among the motives for farmer support for dairy cooperatives. The division of milk producers into sub-groups differentiated according to production technology and scale of operations will be discussed, and the

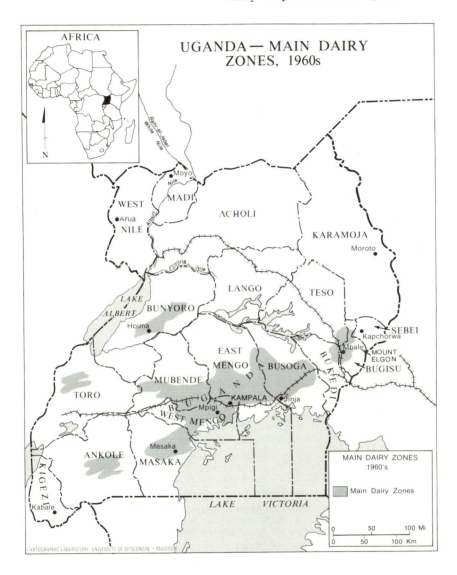

AFRICA

N

UGANDA — MAIN DAIRY
ZONES, 1960s

Bahr-el-Jebel
White Nile

Moyo
Nile

WEST
Arua
NILE

MADI

ACHOLI

Albert Nile

KARAMOJA

Moroto

Victoria Nile

LAKE
ALBERT

BUNYORO

Hoima

LANGO

TESO

SEBEI
Kapchorwa

Mbale

EAST
MENGO

BUSOGA

BUKEDI

MOUNT
ELGON
BUGISU

MUBENDE

TORO

U G A N D A   KAMPALA   Jinja

B U
WEST MENGO   Mpigi

Masaka

ANKOLE

MASAKA

KIGEZI

Kabale

LAKE   VICTORIA

MAIN DAIRY ZONES
1960's

Main Dairy Zones

0   50   100 Mi
0   50   100 Km

CARTOGRAPHIC LABORATORY, UNIVERSITY OF WISCONSIN — MADISON

special characteristics of dairy cooperative officers as an elite group among milk producers will be explained. The goals of different types of society officers will also be analyzed. Finally, we consider the implications of the interaction of differing types of society membership and leadership for cooperative ability to realize the development goals of equality and efficiency.

Having reviewed the motives and actions of society members and officers, we will turn to a review of the ups and downs in governmental backing for cooperatives in the dairy sector. Our analysis reveals attempts to extend government control, at the expense of cooperative autonomy, on the background of an unusual division of administrative responsibility for the supervision of dairy societies' affairs. The failure of these efforts demonstrated the value of cooperatives as an organizational resource, and crucial support in the efficient structuring of a promising new line of farm output.[1]

Realistic evaluation of the relative economic-technical advantages of cooperation interacted with political and bureaucratic ambitions and interagency tensions to determine governmental agencies' approach to dairy cooperatives. Analysis of the shifting treatment accorded dairy cooperatives enables us to demonstrate several points suggested above in Chapter 2 concerning the factors determining the choice of cooperatives as opposed to alternative policy instruments, and the significance of the choices made. Our first objective will be to analyze the motivations for dairy cooperatives. Farmer perspectives are suggested in Table 7.1, based on a survey of dairymen in Bulemezi County, Buganda.[2]

MOTIVATIONS FOR COOPERATIVE MEMBERSHIP

Material Incentives

Cooperative involvement in milk marketing appeared on a significant scale in the mid-1960s. Two main factors encouraged farmer interest in cooperative organization of milk sales. First, the intensive dairy farmer appeared on the scene alongside the traditional extensive stockman. Second, within the framework of efforts to promote milk production and milk marketing in order to eliminate milk imports from Kenya, government began to call for farmer organization into cooperatives as a condition for receipt of various types of aid which it began to make available to milk producers. In addition to supplying societies with cooling equipment, the services of trained milk-handling personnel, and/or transport services gratis or at reduced cost, governmental agencies lent their backing to societies' attempts to gain exclusive control of milk collection and local sales in the trading centers and towns in which they were based.[3]

Table 7.1

MOTIVATIONS FOR COOPERATIVE MEMBERSHIP

Farmer responses to the question:

Why did you join (do you intend to join) a cooperative society of farmers?

	%*
Cooperation is good; good to cooperate; good to work together	7
To serve, help: area, stock farming, the nation	5
Because I am a stock farmer; to associate with stock farmers	6

Percent of farmers giving responses suggesting purposive-solidary motivation: 18%**

In response to government's call	4
To get X from government	4
To have voice; make desires known to government	3
To get X (usually supplied by government, but government is not explicitly mentioned in response; i.e., loans, tractors, advice)	11

Percent of farmers giving responses touching upon government cooperative ties: 21%**

Strength in unity	3
Weakness of the individual	9
Useful; brings benefits; solves problems	11
To get assistance	3
To get X from society	5
To obtain farm supplies through the society	6
To sell, find markets for milk	53

Percent of farmers giving responses suggesting materialistic motivations: 89%**

* N = 141; the percentages total to more than 100 percent because many farmers mentioned more than one reason for joining (expecting to join) a cooperative.

** These percentages have been figured without double-counting; that is, a farmer giving two or more answers within the same broad category is counted here only once.

The breeds of cattle native to Uganda can produce only low yields of milk and meat. Local stock are typically maintained within a system of extensive husbandry; that is, the stock are for the most part expected to fend for themselves, making the best of the natural pasture and crop stubble on which they are grazed. Stock farming does not entail investment in the improvement of stock or of lands. Cash outlays for purchased inputs are minimal: supplementary feeding of purchased foodstuffs is not practiced; and though drugs and vaccinations have in recent years become common items of use, those medicines required for the handling of local stock are typically provided on a highly subsidized basis by the Veterinary Department as a veterinary public health service.

The typical extensive milk producer can thus expect minimal cash expenses per unit of milk produced, but only a very limited milk output per cow. Milk marketing on the part of extensive cattle keepers was typically a matter of sales of a few spare pints to neighbors; or, in the case of owners of larger herds, the house-to-house vending of available surpluses to residents of local trading centers or nearby towns. In addition, produce traders were also to be found collecting milk from a number of herdsmen and transporting it to larger urban centers, where demand for milk was strong.

During the final years of the colonial period and the first years of independence, government began to promote the diffusion of new methods of farm milk production. The limitations which the colonial government had placed upon African farmer access to "exotic" (European) or cross-bred stock were gradually lifted, and replaced by a policy of encouraging selected African farmers to buy exotic cattle.[4] After independence, such programs were expanded and given new emphasis.

The milk production potential of exotic cattle far exceeds that of local stock. However, the acquisition of exotic cattle and the husbandry methods required to guarantee the survival and productivity of these high potential animals demand what are, by the standards of Ugandan agriculture, very great investments in farm improvement. Furthermore, the rearing of exotic stock also demands what are, by Ugandan standards, considerable recurrent outlays of cash for the purchase of production inputs, primarily concentrate feeds, acaricides, and higher cost drugs, such as antibiotics, whose use is justified by the higher value of exotic stock.

Thus, particularly in those ecological zones best suited to the intensive husbandry of exotic stock, there has appeared in the village, alongside the majority of extensive stock keepers, a small but growing core of intensive dairy farmers. For many of the latter, intensive dairy farming is an endeavor entailing substantial cash outlays, undertaken for the explicit purpose of obtaining milk for sale on a profitable basis. Moreover, for a considerable minority, bank loans supply a portion of the capital invested in dairying; in Bulemezi, for example, 28 percent of those with improved stock had received such financial assistance. Thus, the pressure of an external creditor, demanding loan repayment, is added to the internal pressure of the farmer's own desire for profits.[5]

Such profit-oriented, technically sophisticated, intensive dairymen are clearly at a disadvantage in any price competition with stockmen keeping local stock by extensive methods. The latter are able to continue to sell their milk until its price in the market declines to a level equal to the marginal value of additional on-farm milk consumption (plus transport costs), given the minimal cash outlays entailed in extensive milk production. The intensive farmers, by contrast, must obtain a price high enough to enable them to cover their cash outlays and to provide a return on their investment, if they are to achieve their objectives in undertaking a new, advanced husbandry system.

If we take into account, then, that the output of a relatively small number of intensive dairy farms can swamp the milk market of a typical rural trading center, it becomes clear that the great majority of intensive dairy farmers were in need of some means of joining together so as to prevent the bottom from falling out of the local milk market and to organize the cooling and transport of excess supplies to the Dairy Corporation's urban milk handling facilities. Since governmental policy implied the application of persuasion and coercion to induce all local area farmers to join, or at least to market their milk through cooperatives, assistance which could not be expected by a private commercial undertaking, cooperation was obviously to be preferred as the organizational framework for control of local milk marketing and the "export" of surpluses.

Thus, a cooperative monopolistic position within the local market made possible the maintenance of relatively high wholesale and retail milk prices in rural trading centers.[6] The margin on local sales could then be applied to the cost of transporting "surplus" milk, which could not be absorbed by the local market at the monopolistically determined price, to one of the urban processing plants. The margin on local sales also helped to compensate the society for the relatively low price which it received for milk which it sold to the plants. Indeed, cooperative and veterinary department staff, foreign experts, and society officers were all clearly in agreement that the effective exploitation of a higher price local market, so as to "subsidize" low-price, high-cost sales of milk to the urban centers, was one of the primary factors accounting for the degree of commercial success achieved by dairy cooperatives. Not surprisingly, then, in the light of their economic-technical situation and the services which cooperatives could perform for them, intensive dairymen played a leading role in the establishment of dairy cooperatives in their areas and in embryonic attempts to create a national, farmer-based organization for the control of milk marketing on a Uganda-wide basis.[7]

With regard to Buganda, the prime intensive dairying area and the region in which the largest number of dairy cooperatives are to be found, it must be noted that the organization of cooperatives dominated by intensive farming circles is a process in which economic and ethnic lines of differentiation overlap. For most Ganda, local breeds of cattle serve as a side undertaking, a form of saving and investment and a symbol of status and wealth. However, cattle-keeping is viewed as a low status occupation, and as such is often left to members of cattle-keeping ethnic groups from the western districts, or migrants from Rwanda or Burundi. These minority group herdsmen tend the cattle of one or more Ganda agriculturalists, on a milk-and-calf commission or a cash wage basis. Herdsmen may also have cattle of their own which are grazed with those of their employers.[8] In addition to such employees, many westerners are to be found in Buganda engaged in pasturing herds wholly their own. Few of the western herdsmen in Buganda have made the switch from extensive to modern intensive dairying methods; indeed many continue to lead a semi-nomadic existence and do not possess individual rights in land in the areas in which they graze their stock.

In sum, cooperatives served the functions of monopolistic organization of local milk markets and the organization of the long distance shipment of "surplus" milk to urban processing plants. The ability to fulfill these functions was largely dependent upon the encouragement and assistance provided to dairy societies by the various agencies of government dealing with cooperation and milk production. Accordingly, when most dairy farmers questioned in the survey explained the attraction of cooperative membership in terms of the desire to market milk, this orientation must be understood against the background of the favored position which government had assured to cooperatives operating in this field vis à vis any form of private undertaking.

Solidary and Purposive Incentives

Of the farmers who reported membership or a positive attitude toward future membership in dairy cooperatives, 89 percent gave responses which indicated that membership was connected to the pursuit of material objectives, in terms of benefits provided by the society itself and/or in terms of winning government favor and obtaining governmental assistance via cooperation. At the same time, however, a considerable minority of 18 percent made mention of solidary or purposive ends in their responses, expressing generalized support for cooperation (obwegassi: literally, "unity," "uniting")--voicing a desire to advance stock farming, the local area, or the nation as a whole through society membership, or indicating that the attraction of cooperation was a function of self-identification as a stockman. Similar references (service to profession, local area, or nation) may be found in a good number of society officers' descriptions of their cooperatives' objectives.[9] This suggests that the cooperative was understood by a considerable minority of its members as something more than a purely utilitarian material device.

This solidary-purposive aspect was not strong enough to sustain society operations where a cooperative proved unable to provide a minimum level of material incentives--regular payment for milk at a price considered reasonable by members and nonmember suppliers. However, it did mean that the cooperative functioning with a reasonable degree of commercial success was an organization possessing broader significance for its members than the mere provision of an outlet for their produce.

As will be noted below, the leadership of the dairy societies includes individuals involved in a variety of public activities, including crop cooperatives and religious, school, and other community organizations. The mobilization of society members' support for positions on issues of local concern may, then, come to serve as a resource in the hands of such leaders, given their own and some members' understanding of the cooperative as an organization dedicated in part to the promotion of community welfare.[10]

DAIRY COOPERATIVES IN INTENSIVE DAIRYING AREAS: LEADER BACKGROUND CHARACTERISTICS AND MOTIVES

In this section we make use of our survey data to compare the officers of Bulemezi's three major dairy cooperatives with the county's other dairy farmers. The picture of society officeholders which emerges suggests a number of perspectives, complementary to those already presented by researchers dealing with rural community and cooperative leadership in Uganda.

We find, first of all, that 21 out of 22 officeholders were intensive dairymen of Ganda ethnic origin (and the one exception played a negligible role in the affairs of his society). Extensive milk producers and members of the western minority ethnic groups were not represented among the societies' officers, nor indeed within the societies' registered membership, in proportion to their importance as suppliers of milk to the societies (Table 7.2). The wealth, economic-technical competence, and initiative which brought intensive dairymen to innovative endeavor, together with their membership in the locally dominant ethnic group, enabled intensive dairymen to control the leadership of milk marketing societies.

Table 7.2

CONTRIBUTIONS* OF NON-GANDA, EXTENSIVE PRODUCERS TO THE MILK
INTAKES OF THE THREE MAJOR BULEMEZI DAIRY COOPERATIVES

	Society A (%)	Society B (%)	Society C (%)
Milk supplied by nonmember, non-Ganda extensive producers, as % of total milk intake	13	18	39
Milk supplied by all non-Ganda, extensive producers (members and nonmembers), as % of total milk intake	22	20	66
All non-Ganda, extensive producers (members and nonmembers), as % of total number of farmers supplying milk	40	35	62

   * Figures given are nonweighted means for the months of January, March, and May 1973.

     The need to direct the operations of the dairy cooperatives
is a function of the economic demands of intensive dairying.
Intensive dairying, as noted above, requires large investments,
while possessing a high profit potential. Realization of this
potential and achievement of an adequate return upon investment
require that the farmer have a ready outlet through which he can
dispose of his daily milk production at a reasonable price. Both
those seeking to be leaders and those electing them were apparently influenced by such considerations in the societies being
discussed: leadership positions fell in most cases to those for
whom society office was a vital complement in the field of marketing to effort and capital invested in intensive dairying.
     Thus, as will be demonstrated, leadership was not a direct
function of absolute levels of income, but rather was related to
the absolute level and relative importance of dairying and farming in general in the economic activities of the society officers. Furthermore, society leadership would not seem to have
functioned here as an instrument of mobility strategies, aimed
at helping the society officer to gain entrance into regional or
national arenas of social, economic, and political activity.
Instead, leadership is a function of the importance of rurally
based agricultural activities. Indeed, it will be seen that re-

sources gained through the success of past mobility efforts were
activated in order to contribute to the success of society leadership, rather than vice versa. Finally, we will suggest that
dairying, as an element in the acquisition of wealth, power, and
status within the rural areas themselves, contributed to the
growth of a leadership group, modernist in terms of sources of
its status and its economic successes and yet retaining an orientation to the local community as a primary focus for economic,
social, and political activity.

Tables 7.3 and 7.4 compare various aspects of society officers' farm and overall economic activities against those of other
dairy farmers in Bulemezi. These tables show that the major distinguishing characteristic of the society officers is their status as outstanding farmers. It is the extent of their operations
which sets them apart, whether measured in terms of total farm
profits, milk sales, investment in intensive dairying, total herd
size, or percent of improved breeds within the dairy farm herd.

It is interesting to note that the proportion of officers
with no nonfarm income is equivalent to that of the other farmers, and that officers do not clearly outstrip nonofficers with
regard to nonfarm income, as they do in farm profits (Table 7.3).
Indeed, in the highest nonfarm income bracket, it is the nonofficers who predominate. Several factors account for this finding.
First, in many cases dairying is a marginal undertaking for those
with the greatest nonfarm income. On the other hand, where
dairying has been undertaken on a substantial and serious basis,
the importance attached to the farm and its size frequently jus-
tifies the employment of a full-time farm manager. Furthermore,
the nonfarm occupations of those with sizable incomes from those
occupations often require that they work, or even reside, in
areas removed from their dairy farms. Thus, for dairy farmers
with the highest nonfarm income, nonfarm careers typically constitute the primary focus of interest and direct personal involvement; the dairy farm is of secondary importance. Society
officers, by contrast, are usually individuals for whom dairying
is of greater relative importance, individuals who therefore as-
sign higher priority to milk marketing activities--such as the
pursuit of coopererative office.[11]

The desire to use cooperative office as a stepping stone for
advancing to nonfarm occupations in regional or national centers
of economic and political activity would not seem to have played
a significant role in bringing individuals to office in the three
major societies. Intensive dairymen were the predominant element
within the active membership of each of the societies. Indeed,
their stake in efficient cooperative performance was so great
that they could hardly permit society leadership to be dominated
by individuals or groups for whom successful society operations
were not important. In the majority of cases a substantial and
central involvement in farming in general and dairying in partic-

Table 7-3

TOTAL FARM PROFITS, NONFARM INCOMES, AND TOTAL INCOME (PREVIOUS YEAR):
DAIRY COOPERATIVE OFFICERS* AS COMPARED TO OTHER DAIRY FARMERS**

Total Farm Profits	Officers (%)	Others (%)	Nonfarm Incomes	Officers (%)	Others (%)	Total Incomes	Officers (%)	Others (%)
Less than 2,500 shillings	16	63	None	32	31	Less than 2,500 shillings	0	36
2,500 shillings-5,000 shillings	26	21	Less than 2,500 shillings	21	26	2,500 shillings-5,000 shillings	21	15
5,000 shillings-7,500 shillings	11	7	2,500 shillings-5,000 shillings	16	7	5,000 shillings-7,500 shillings	21	11
More than 7,500 shillings	42	7	5,000 shillings-7,500 shillings	21	8	7,500 shillings-12,500 shillings	21	11
Not available	5	2	More than 7,500 shillings	5	26	More than 12,500 shillings	32	23
			Not available	5	2	Not available	5	4

(continued)

(Table 7-3 cont.)

* N = 19. Three officers are not included:
(a) Vice-chairman, Society B: This farmer refused to be interviewed. His intensive dairy operation was small, and he did not
have milk to sell to the society every month. He was very
wealthy, owner of a Kampala-based transport firm and of the
thriving bar in trading center B.
(b) Treasurer, Society C: This farmer was not included in the
survey as his farm and home were not located in Bulemezi, but in
neighboring Buruli County. He was the biggest milk supplier to
the society among the members. He had nonfarm economic interests, including transport services which he supplied with his
pickup truck.
(c) Committee member, Society C: This farmer was a non-Muganda
extensive producer. He was not active in society affairs, and
four out of five fellow officers surveyed could not recall his
name when asked to list the society's officers. He had not sold
milk to the society for some time, and in fact had taken his herd
to Buruli County in search of better pasture and was not to be
found in trading center C.

** N = 121. Two recipients of confiscated Asian estates are
not included.

ular was necessary as a preliminary indication of bona fide concern for the society's success in order to make possible election
to office.
     Moreover, if we turn to society officers with a substantial
involvement in dairying, we find that most were not at a point
in their life cycle which suggests aspirations to rise up and
out of the village. To the contrary, 9 out of 19 were older men
(over 45), for whom dairy operations were a climatic stage in
active economic careers in which most had earlier pursued occupations which drew them to the city or to distant parts of rural
Uganda. Investment of previous savings in dairying was intended
to provide the individual with an adequate income from his own
home farm in the latter years of his life.
     Even for the younger officers, society leadership would seem
to have been functionally related to a substantial economic involvement in milk production; two cases may be cited in illustration of this point. In one instance, a young man, part owner of
a large farm together with a number of other members of his immediate family and an employee in his father's wholesale business
(which was located in the same trading center as the society's
headquarters, a few miles from the farm), resigned his office as
treasurer of the society in order to take a professional course

Table 7.4

CHARACTERISTICS OF DAIRY FARMS:
DAIRY COOPERATIVE OFFICERS AS COMPARED TO OTHER DAIRY FARMERS

	Officers (%)	Others (%)
A) Years since first acquisition of exotic stock:		
1-4 years	21	62
5 or more years	79	40
	N = 19[a]	N = 102[b]
B) Investment in intensive dairying:		
Less than 10,000 shillings	10	61
More than 10,000 shillings	90	39
	N = 19	N = 102
C) Herd size:		
1-15 cattle	31	72
16 or more cattle	70	28
	N = 19	N = 123[c]
D) Exotic stock as proportion of total herd:		
0-50 percent	21	50
51-100%	79	50
	N = 19	N = 123
E) Value of milk sales (previous year):		
Less than 4,000 shillings	27	79
More than 4,000 shillings	69	20
Not available	5	1
	N = 19	N = 121[d]

[a] Three officers are not included; see notes to Table 7.3.
[b] Two recipients of confiscated Asian estates and 19 farmers without improved stock on their farms are not included.
[c] All nonofficers are included.
[d] Two recipients of confiscated Asian estates are excluded.

connected with his business career. Was cooperative office then merely a stepping stone, a means of obtaining experience, making contacts, and facilitating progress onward to more interesting and rewarding positions? The continuation of the story suggests that it was not: the former treasurer was later returned to office as vice-chairman and was active in society affairs, despite frequent trips on behalf of his father's wholesale agency. Moreover, in the course of his interview, he mused over the possibility of planting sugarcane on the family farm and concentrating his efforts to a greater extent in agriculture.

The second case is that of the secretary of another of the societies. Toward the conclusion of our field work, this individual decided upon the liquidation of his dairy farm in order to concentrate his energies upon his transport company in Kampala, and thus began to sell off his cattle. Though he had not yet resigned, we may assume that leaving office in the cooperative stood as a natural complement to this decision. Here again, however, the behavior of the society secretary cannot be understood in terms of the use of cooperative office to promote personal mobility. Rather, his dairying and cooperative activity represented a revival of economic involvement in his rural home area after a number of years of employment in the city. Indeed, rather than the cooperative serving the officer's individual ambitions to create contacts in regional and national centers, it would be more appropriate to say that here the officer's previously acquired familiarity with persons and processes in the center was mobilized on behalf of the cooperative. He applied his experience and contacts in Kampala to the advancement of his society's interests. Largely through his initiative and persistent lobbying the society achieved two prizes, which were rarely granted at the time: official registration as a cooperative society (granted by the Co-operative Department), and the installation of a bulk milk cooler (provided by the Dairy Corporation). When the secretary decided to abandon dairying, this came largely in response to unexpected new opportunities in transport which were created by Amin's expulsion of the Asians, rather than as part of the realization of a plan for personal social and economic advance in which cooperative office played a merely instrumental role.

The society officers, then, were farmers, and usually dairymen, of substance; not the absolute level of their wealth brought them to cooperative office, but rather the extent and centrality of their involvement in farming and dairying. In sum, we find in these examples the beginnings of a modernist, professional farm leadership. This impression is strengthened when we note that some two-thirds of the dairy cooperative officers had also held office in some additional cooperative (as opposed to only one-third of the general dairy farming public).

Five of the 19 society officers remind us of the "man of profit," as described by Mafeje in his study of commercial farmers in Busiro County, West Buganda District.[12] In response to queries concerning general leadership status in the local area, these individuals denied both being or wanting to be community leaders. On the other hand, in three out of five instances these officers did report having served as officeholders in other cooperatives in addition to the dairy society. These, then, were men for whom organizational involvement and acceptance of leadership were integrated with the pursuit of individual farm-economic concerns. Such men were not interested in taking responsibility for public affairs which did not directly affect their economic undertakings, since such responsibility would tend to detract from their ability to concentrate time and energy upon personal enterprises.

On the other hand, 13 (68 percent) of the officers did feel that they were local leaders, and 7 among them indicated that this situation was in keeping with their wishes. Eight of the 13 were members of various local social organizations (church and church-sponsored groups, parents' associations and school boards), including three of the six leaders who denied aspiring to leadership. (By contrast, none of the five "men of profit" reported such organizational activity.) These men acknowledging local leadership, interest in community leadership, and/or active participation in community organizations bear some resemblance to Mafeje's "men of affairs." As described by Mafeje, "men of affairs" are usually related by descent or social ties to the "chiefly establishment" of Buganda and often the patron-landlords of a large number of tenants. They fill local leadership roles whose roots lie in the patterns established by the Kabaka, his court, and his appointed chiefs during the historical evolution of the Kingdom of Buganda. Individuals corresponding to this prototype ascribe greater importance to superior social and political status and the quality of their personal relations with their neighbors/chiefs than to economic self-aggrandizement, narrowly conceived.

In almost half (6 out of 13) of the cases, those evincing interest in being leaders or reporting current involvement in community affairs did possess some tie in their family or personal background which linked them to the chiefly authority structure of Buganda (four sons of chiefs, including one who was himself a chief; one son of a sub-county clerk; one retired chief). However, most of those in the group were of commoner background and had not themselves served as chiefs or officials filling related administrative positions. In addition, only one individual (the son of a chief, himself a chief) in the entire group had any significant number of tenants (8) on his land).

While it is clear, then, that these individuals differ from the "men of profit" in their involvement in community affairs, it is not clear whether they can all be equated with the "men of affairs" in terms of the nature of their authority and involvement in community life. To the contrary, some clearly based a prominent position in the local community upon a foundation of individual professional and economic success, rather than on ties to the chiefly hierarchy or the building of a personal retinue of tenant-clients. Thus, although their numbers are limited, we may suggest that the community-oriented officers without chiefly ties (and possibly some of those with chiefly ties as well) represent not only a professional farmer leadership, but a new form of community leadership, basing itself upon economic and professional success in the cash economy and expressing itself through participation in and direction of the new formal organizations of the local community.

The appearance of such leadership in the local community results from the introduction of innovations, such as intensive dairying, which narrow the gap between the technical sophistication and income possibilities of agricultural (or combined agricultural and commercial or salaried) activity in the rural areas and the challenges and rewards of urban careers. Where processes of agricultural and overall socioeconomic change in the rural areas have begun to create opportunities equivalent to those available to the middle and upper-middle strata of urban society, then even a young, talented, and educated individual can calmly decide to make the local community his primary arena of activity without feeling that such a choice entails acceptance of inferior social status or a lower standard of living.

As for the overall involvement in community affairs to which these men incline, it is far from clear that they will profit less in the long run from the allocation of time and energy to leadership in community affairs, as opposed to wholehearted devotion to matters of immediate economic concern. Prominent standing, status, and power acquired in the community may well provide dominant individuals with advantages in private economic dealings and help them to achieve favored access to the economic benefits of governmental programs, as well.

SOCIETY OPERATIONS

We may now turn to a more broadly based review of modes of cooperative operation, seen as a function of the interaction of leadership motivations and the economic-technical conditions facing the societies. What are the basic objectives of leaders of the type described above in their management of society affairs? What particular means and subgoals may be adopted in keeping with the special circumstances within which each cooperative functions? What is the nature of leadership and society operations in those cases mentioned above, in which society leaders were not intensive dairymen and were not personally engaged in the marketing of milk via their societies? To answer these questions, we will examine the activities of a number of the Bulemezi cooperatives and will also consider an interesting example drawn from a neighboring county in East Buganda District.[13]

The Katikamu Dairy Farmers Cooperative Society Limited in Bulemezi represents the prototype of the dairy cooperative founded by a core group of serious dairymen in order to guarantee an outlet for their milk and the profitability of their investment in dairying. The origins of the society lie in an association of farmers established to advance the work of the artificial inseminator stationed in Wobulenzi, the trading center immedi-

ately adjacent to Katikamu, a subcounty headquarters. By virtue
of this precedent of organized activity in conjunction with government departments, the establishment of a number of intensive
dairy farms in the area, and a favorable location (some 30 miles
north of Kampala on the main road), Katikamu farmers were in 1966
among the first to be provided with a milk cooler, supplied to
the Uganda government by UNICEF. Later, the Dairy Corporation
provided a bulk tank cooler of 500 gallon capacity, and the society hoped to be allocated a new and bigger device.

The Society's association with governmentally backed milk
production and marketing efforts enabled it to claim control of
the milk trade in Wobulenzi and the surrounding milkshed and to
demand the intervention of chiefs and the representatives of central government ministries to enforce this monopoly. The Society
did in fact succeed in dominating the local milk market and
achieved a considerable degree of success in its marketing operations. Despite rising milk input, the Society proved able to
dispose of an increasing percentage of its intake through local
and other high-return outlets. Though the officers as a group
did not demonstrate outstanding initiative (neglecting the possibility of marketing Society milk in Bombo, a large regional center ten miles south of Wobulenzi on the Kampala road) and were
aided to some extent by coincidence (one farmer-member was a food
wholesaler and purchased a large quantity of milk from the Society daily to aid him in meeting his commitments to institutional
customers in Kampala), they did display sufficient basic competence and honesty to enable the cooperative to exploit its local
monopoly effectively and prosper. In the early years of Society
operations, the officers were moderate in their payment of allowances to themselves from Society funds. They raised this form
of reward to significant levels only in later years, by which
time Society operations had considerably expanded and been put
upon a soundly profitable basis.

Two factors suggest themselves as explanations for this
prudent strategy: on the one hand, the considerable depth of involvement of the officers in dairying and, on the other hand, the
fact that facing the officers stood an additional core group of
intensive dairy farmers who were Society members. All but one
of the officers were involved in regular milk sales to the Society; and the exception, the Society's former secretary and current secretary manager, was building an intensive dairy operation
as one of the main planned sources of income for his later years.
Indeed, the average per officer milk deliveries in the Katikamu
Society were by far the largest in the country.

Exploitation of society resources for quick and easy gains
in the short run, then, threatened to revenge itself upon the
officers by jeopardizing, in the medium to long run, the market
outlet for their milk. This, in turn, would jeopardize the profitability of the officers' investment in dairying. As noted

above, for such men cooperative leadership was a natural and integral aspect of their involvement in dairying. Our findings regarding the Katikamu Society suggest that cooperative office--i.e., the management of milk marketing--drew a managerial effort in keeping with the importance of dairying in the officers' individual economic undertakings.

As for the second factor, Society officers faced a membership which contained a significant group of concerned peers: well educated, high income farmer-members who also had made considerable commitments in dairying and thus had a major economic interest in sound marketing operations. These farmers were in no sense clients of the officers. They were both free and capable of voicing dissatisfaction with the operation of Society affairs, were officers' behavior to harm their interests as dairymen and Society members. In the Society under discussion, generally successful management meant that this capability remained latent, but in several other cooperatives we noted complaints and challenges to leadership raised by dissatisfied peers among the societies' members.

To supplement this picture of the Society led by dairymen, in which Society activities can be adequately understood in terms of the need to market officers' and members' milk, we present the example of two other societies. In each cooperative, a core of intensive dairymen were to be found for whom milk marketing was a vital adjunct to farm milk production, as in Katikamu. However, additional factors operated in these other cases which led officers to adopt special measures of an interesting nature as part of their efforts to guarantee societies' ability to succeed in their task of milk marketing.

The Semuto Dairy Farmers Cooperative Society stands upon the foundation of a group of intensive dairymen from whom it draws its leaders. As in the case of Katikamu, the Cooperative served as the means by which intensive dairymen claimed the right to monopolize the local milk market with governmental backing. The Cooperative also provided the framework for a successful lobbying campaign through which the farmers obtained a bulk milk cooler, an essential tool in both local marketing and the shipment of surplus milk to Kampala.

The Cooperative's leaders were acutely aware of the need to dispose of a maximal percentage of its milk in the local area, so as to guarantee Society profitability. Thus, they devoted considerable effort to the development of a network of sales agents in Semuto and surrounding trading centers. Despite these efforts, the continuing need to dispose of a sizable percentage of Society milk to low price outlets and the transport costs associated with milk marketing caused the Cooperative considerable financial problems.

To remedy their plight the officers took an interesting, and revealing, step: they decided to enter the field of long-distance milk trading. The Cooperative began to collect milk from nonmembers at Wakyato, a trading center lying to the north in an extensive grazing area. It was hoped that profits from this milk, which was purchased at a price below that paid at Semuto itself, would boost revenues, and that the addition of considerable quantities of milk to the Cooperative's intake would help make possible more economic transport of milk to distant markets. (In fact, the cost of transport from Wakyato to Semuto proved prohibitive, given the quanitities which could be collected there, and the experiment caused additional financial difficulties to the Cooperative. The project was not stopped, however; the Cooperative's president took over the Wakyato milk trade as a personal venture, on the assumption that it would eventually prove profitable.)

This episode in the Semuto Cooperative's operations provides an extreme and obvious demonstration of a phenomenon which exists in virtually all of the societies operating a milk collecting center: cooperative commerce in nonmembers' produce. As noted above, in the three major Bulemezi cooperatives we find societies, dominated by Ganda intensive dairymen, which buy a substantial percentage of their milk intake form nonmember extensive herdsmen.

Acceptance of nonmembers' milk is in the first instance an obligation imposed upon the societies by the dairy agencies[14] in return for the various forms of material assistance and the monopoly privilege granted the societies. Particularly prior to the coup, the leadership of these agencies demonstrated considerable sensitivity to the possibility that cooperatives might depress the total quantity of milk entering organized marketing channels by excluding the milk of those not belonging to the societies, or otherwise discriminating against nonmembers.[15]

In addition, in certain circumstances the acceptance of nonmembers' milk may be economically advantageous to the society. Where local demand is strong relative to members' ability to expand production, the availability of milk from nonmembers means the possibility of increasing sales and profits without increasing the number of farmers who will share in the distribution of profits, or take part in the internal affairs of the cooperative. Alternatively, where society production itself exceeds local demand at the price desired, or where a requirement to buy nonmembers' milk means that a surplus cannot be avoided, it will be in the interest of the society to obtain maximal quantities of nonmember milk if milk transport is paid for at a fixed rate per delivery. Here, raising the surplus over local demand spreads the overhead expenditure on transport over a larger quantity of produce, while bringing in additional sales income.

By way of contrast, the Bugerere Balunzi Dairy Farmers' Cooperative Society illustrated conditions in which the membership of those selling milk to the society is seen as positive and necessary. This Cooperative's officers sought to establish for their Society an extensive county-wide monopoly, rather than simply controlling a single trading area and its vicinity. In order to realize this ambition they were in need of large amounts of capital, which eventually led them to convert milk suppliers to Society members.

The Bugerere Balunzi Society also possessed a core group of intensive dairy farmers. Its headquarters site in Bukoloto was a natural collecting point for the milk of not only the immediate hinterland, but rather the entire length of Bugerere County, extending northward. Considerable milk from the north, which lacks any sizable population centers, can be marketed in the southern area, and the rest must pass through the south on its way to Kampala. The society began operation on a county-wide basis from the start, with the help of Dairy Corporation transport and cooler.

Operations on such a scale led to relatively large investments, primarily for the purchase of milk transport vehicles after the withdrawal of those provided by the Dairy Corporation. This meant substantial debt burden and heavy maintenance and operation expenditures. Effective control over vehicle use was difficult; even some officers used them for private ends. High staff costs also plagued Society operations.

Thus, despite respectable performance by the Society leaders on a number of points (a successful local marketing campaign; a reasonable self-restraint in the drawing of various allowances from the Society treasury), the Society's current revenues proved inadequate to meet loan obligations and current expenditures. To postpone the inevitable day of reckoning with the over-extension of Society operations, officers sought to raise needed funds by forced-draft recruitment to Society ranks of all those making use of the Cooperative's marketing services. New members were naturally required to purchase membership shares; in addition, special levies, credited to member share accounts, were imposed upon member milk payments. Payment of a differential, lower price to distant nonmember suppliers, as in the Semuto case, was ruled out by the precedent of a uniform, county-wide price established during the period of the Dairy Corporation's involvement in milk transport and the possible reaction of the Corporation and the affected farmers to violation of this long-standing practice.

The introduction of large numbers of farmers into the Society eroded somewhat the monopoly of control over Society affairs held by Ganda intensive dairymen. A number of non-Ganda had served as committeemen from the early years of Society operations. One such individual was coopted to the Society's leadership after helping to organize a northern protest against aspects of Society functioning. This non-Ganda served for many years as the Society's treasurer and toward the end of the period of our research was elected vice-chairman of the Society.

The non-Ganda treasurer was, however, isolated within the leadership group of the Bugerere Balunzi Society. In a joint interview from which he was absent, other officers spoke scornfully of the election of representatives from the north as a political concession to ethnic sentiment, which had required that criteria of capability be overlooked. The gesture to the members in the northern areas was apparently considered a reasonable price to pay for the recruitment of share funds from a large number of farmers for the maintenance of the Society's extensive operations.

Thus far we have spoken of societies in which the dominant leadership element is composed of farmer-members, marketing their milk output through their cooperatives. Societies were found in which officers were neither intensive dairy farmers, nor suppliers of milk produced by local stock. For these officers, commercial and financial interests, rather than the need to market agricultural produce, were the principal motives for choice of the cooperative framework.

The organization of milk marketing in Bamunanika provides an extreme case of this phenomenon. Here a number of individuals who did not themselves market milk through the "society" managed to channel milk supply in the area through their hands, under the guise of cooperation. The Bulungi Bwabasumba Society of Bamunanika operated in a shop immediately adjacent to the general store of the Society treasurer, who supervised the conduct of Society operations. Each month some 40-45 farmers, none of them keeping exotic or crossbred stock, delivered milk. According to the treasurer, a Ganda, a required deduction of 5 shillings from the payments due those beginning to sell milk to the Society converted suppliers into members. From month to month, between 45 to 60 percent of the members, accounting for roughly equivalent percentages of the total milk intake, were non-Ganda.

The case of the Kimwanyi Dairy Cooperative Society is more complex than that of Bamunanika. The origins of the Kimwanyi Society lay in the desire of a number of individuals to develop 640 acres of land owned by one of them as an intensive dairy farm. Involvement in the farm was to take the form of investment of capital; the actual work of farm development and operation was to be left to hired workers. The Society's secretary explained the founders' motives quite frankly: the local Veterinary Department staff member had urged them to organize as a cooperative in order to qualify for special government assistance available to cooperatives only.[16]

The founders hoped that cooperative status would help them to obtain both farm development loans and government aid in attempts to dominate the local milk market. Indeed, the desire to establish the cooperatively owned farm was closely related to the belief that a large, stable milk market would soon develop in the area. A government hospital had been built at Nakaseke,

the trading center in which the Kimwanyi Society was based, and
its opening was expected to bring a strong demand for milk by
both staff and patients. As cooperative officers, but not as
heads of a private company, Kimwanyi's leaders would be able to
claim monopolistic rights to this milk market. Thus, the milk
trade in Nakaseke fell under the control of a group of Society
leaders for whom milk marketing activity was an instrument in a
campaign to achieve long-range investment goals, rather than a
necessary adjunct to current individual involvement in farm milk
production.

FARMERS, LEADERS, AND DAIRY COOPERATIVES: CONCLUSIONS

   The data presented in this chapter have shown us something
of the general dairy farmer motivation for cooperation. We have
seen the attractions of cooperative leadership to active intensive dairymen, on the one hand, and to traders and investors, on
the other. Of course, in addition to his direct interest in
guaranteeing a good, stable market for his milk, the dairyman-
officer may also have interests in society leadership which overlap those of the trader or capital-mobilizing investor.
   What is striking throughout is how distant these motivations are from the anti-capitalist ethos of the Rochdale pioneers. Dairy cooperation is quintessentially capitalist in inspiration. It is a mechanism for the nascent class of commercial farmers to save, accumulate, and invest. Much more directly
than in the cotton and coffon cases, the dairy cooperative is an
agency of rural capitalism.
   We may conclude our discussion of farmers' involvement in
dairy cooperatives by attempting to evaluate the examples discussed above in terms of the two main developmental criteria of
efficiency and equality. With regard to the first standard, we
have noted that in the zones where intensive dairying is prevalent, commercial dairymen tend to be the dominant element within
society leadership. Officers' own concerns for a stable and
guaranteed milk market, and the pressure exerted by an aware and
active constituency of other intensive dairymen, mean that both
internal and external pressures tend to induce a reasonable standard of management of society affairs.
   On the other hand, where the relative weight of intensive
dairymen among milk suppliers is diminished by a large group of
extensive milk producers, particularly by herdsmen of lower status and immigrant ethnic origin, the rise and functioning of society leadership is influenced to a greater degree by factors not
directly related to the interlinked activities of milk production
and marketing. Our findings suggest that the extensive herders
are more vulnerable to domination by individuals from outside
the "profession," who exploit general status and economic re-

sources to achieve control within the specialized field of milk
marketing.

The implications of this phenomenon for the successul conduct of society operations are ambiguous. On the one hand, even
where those who gain dominant position are themselves intensive
dairymen, the possibilities of personal benefit through exploitation of society resources may tempt officers to irresponsible,
if not corrupt, steps. The temptation to exploit the perquisites
of office irresponsibly seems to be relatively greater to the
extent that the pressure exerted by intensive dairymen's concern
for efficient milk marketing is diluted by the presence of a
strong element of extensive and minority group milk producers in
the society's membership.

On the other hand, we must also take into account the possibility of quiet and competent management of milk marketing by
traders for whom the cooperative is an important extension of
their commercial undertakings. (The case of Bamunanika is relevant here.) What will nonetheless be lacking in such cases is
the active role of farmers and farmer-leaders in the management
of society affairs and the participation of farmers in the profits of the milk trade. Profits will be appropriated by trader-
officers by one means or another, though without undermining the
society's ability to survive. Thus, while milk marketing may be
conducted efficiently, trader dominance will tend to limit the
cooperative's ability to involve farmers actively in an organization which links them to national and regional centers and provides them with maximal incentives and aid in farm development.

With regard to equality, our findings raise serious doubts
concerning the degree to which cooperation is likely to lead to
greater equality in the dairy sector. In those societies dominated by nonproducer interests the situation is clear: a position
of dominance, based upon general status and economic resources
not directly associated with farming activities, finds expression and is reinforced by control of the organizational tool of
cooperation.

Furthermore, we have noted that even in producer-led societies there arises the vexing question of relations between intensive farmers of the majority ethnic group and extensive herdsmen, many of them members of a low status minority. Thus, for
example, one of Semuto's most active officers spoke to us with
glee of the element of sharp dealing in the Society's trade in
nonmember milk collected at Wakyato. In the Katikamu Cooperative, Society leaders showed indifference to the question of membership of even permanently resident Ganda milk suppliers. The
officers failed to urge a number of such suppliers who were
partly paid-up members to complete their payments and thus protect their right to share in the distribution of bonus. Lacking
a conscious commitment to the principle of equality, officers
failed to take the steps which would have been necessary to over-

come the absence of direct social and economic ties between them
and the non-Ganda suppliers and bring about the registration of
the latter as Society members.

Moreover, problems relating to equality may arise not only
where extensive and minority group stockmen do not become society
members, but also where for various reasons measures are taken
to guarantee their enlistment in society ranks. In the Bugerere
Balunzi Society, for example, an activist, Ganda-dominated leadership imposed contributions of capital to Society funds upon a
largely passive membership, which contained a high percentage of
non-Ganda herdsmen, as well as Ganda breeders of local stock.
Milk suppliers had to affiliate with the Society and contribute
to its capital stock if they wished to make use of its marketing
services. Were parallel steps to be taken by a cartel of milk
traders, they would certainly be viewed as exploitation of monopolistic powers by a privileged group. It would seem that the
steps taken in Bugerere in this field are substantively equiva
lent to such a situation, despite the formally democratic nature
of decision-making processes in the Society.

In sum, these examples tend to confirm our contention that
adherence to an abstract, universalistic concept of equality is
not a likely concomitant of the occupational, ethnic, or community solidarity which may find expression through cooperative
organizational forms. To the contrary, community and ethnic solidarity can coexist with acceptance of economic differentiation
among the bearers of a common identity; while differences in the
scope and sophistication of farm operations weaken the sense of
shared occupational identity and interests. Moreover, while
community and ethnic sentiment may unite some agriculturalists,
they divide them from farmers from other villages and differing
ethnic background.

National political leadership, adhering to egalitarian principles or seeking political gain through the promotion of equality, may attempt to counter those forces working against the realization of cooperation's egalitarian potential. However, in
the case of Uganda's dairy cooperatives, no such outside guidance
was applied in a consistent and effective manner.

WHY COOPERATIVES? DAIRY COOPERATIVES IN THE TANGLED NETWORK
OF POLITICAL AND BUREAUCRATIC INTERESTS

The Ugandan political elite has sought, as a vital goal of
public policy, to create channels of influence through which it
would be possible to direct the agricultural economy, while creating an awareness of government's power. Governmental encouragement of dairy cooperatives may be seen, in the first instance,
as part of this effort to create vital, meaningful links between
national political leadership and the village community, while

also providing an efficient mechanism for promotion of a new productive sector.

The role of cooperation in dairying, however, was determined not only by this general drive to create stable relations of dependence and support between the regime and the agricultural population, but also by the particular interests of certain political figures and administrative agencies. Responsibility for dairy societies was divided between the Cooperative Department and the agencies dealing with milk production and marketing. Thus, the desire to "claim title" to access to a specific clientele group, so as to strengthen positions within the national political elite and the network of governmental administrative units, became a critical factor affecting cooperative-government relations in the dairy sector. While the Cooperative Department has always favored cooperation in the dairy sector in principle,[17] the attitudes of the dairy agencies have shifted in keeping with changing political and economic-technical circumstances. We analyze these circumstances as they affected society-government relations within three periods: 1964-68; 1968-coup; post-coup.[18]

1964-68

As the dairy agencies first elaborated their plans to equip a network of milk collection centers with coolers, they announced their intention that cooperatives were to manage these new dairies; and, indeed, registered or unregistered societies were associated with most of the early milk collection points. This was a logical outgrowth of the enthusiastic government support for agricultural cooperation in the post-independence years. Without some special justification for such a step, it would have been unusual for the Ministry of Animal Industry to take upon itself from the start direct administrative responsibility for local level milk marketing. Furthermore, the Dairy Corporation, established in 1967 to provide the ministry with an appropriate organizational means of involvement in milk marketing and processing, was in its early days ill equipped to assume nationwide milk collection responsibilities.

Thus, until the Dairy Corporation could train staff and gain experience, involving cooperatives in milk collection lessened the administrative burdens in undertaking a new and unfamiliar set of activities. Cooperatives and their officers helped to organize farmers and encouraged them to make use of the new marketing facilities. Where funds were not available for the construction of buildings, the societies might themselves undertake the task, or would locate and rent a suitable structure. Societies hired and supervised laborers and, in some cases, sales and bookkeeping staff. The cooperatives arranged the purchase of minor items required in the day-to-day operation of their centers,

such as detergents, stationery, and milk packaging materials.
Societies also bore the burden of arranging the payment of individual farmers, who at some centers might number over 100; the
Dairy Corporation had only to provide a single check to the society itself for milk accepted at its processing plants.

1968—Coup

As early as 1966, J. Babiiha, the Minister of Animal Industry, had publicly expressed his reservations concerning the efficiency of cooperative management of milk collection and marketing.[19]   By 1968, the second year of the Dairy Corporation's
existence, an open rift had developed between the Corporation on
the one hand and the cooperative societies and the Cooperative
Department on the other. At a number of locations where regis
tered and unregistered societies had been placed in charge of
centers, the Dairy Corporation seized direct control of milk collection, while at many locations Corporation administration was
fixed as the rule from the start. Thus, according to a survey
conducted in the first half of 1972, the Corporation was operating some 43 centers, while only 16 were in the hands of cooperative societies.[20]   Furthermore, where cooperatives continued
to control centers, the dairy agencies sought to curtail the support which they had originally been providing to the societies.
This change in policy was facilitated by improvements in the
technical capacity of the Dairy Corporation. The Corporation had
gained a degree of operational experience and had begun to absorb
the graduates of milk collection and handling courses. Moreover,
from the political point of view, the shift in treatment of dairy
cooperatives formed part of a generally tougher and more critical approach toward cooperation which began to become evident
from 1966 onward. Whereas dissatisfaction with the functioning
of cooperatives in other branches of agriculture found expression
in the appointment of Cooperative Department officials to serve
as supervising managers of major crop unions, in the dairy sector
the extension of government's bureaucratic controls led to the
displacement of cooperative structures and not simply to the erosion of their autonomy.
To understand how this difference in the treatment of cooperatives in the different sectors came about, we must take into
account that political considerations were dominant in the dairy
agencies' decision to undertake a rapid, nationwide, and economically risky expansion of the milk collection network. Milk
reaching the Dairy Corporation's processing plants from the rural collection centers made possible the displacement of imports
from Kenya. Since resentment against Kenya's economic dominance
within East Africa was a major theme of Ugandan nationalism,
this substitution of local production for imports was a political
achievement of considerable significance.[21]   In addition, pay-

ments to farmers for their milk served as the foundation for
ties of substance between the dairy agencies and a growing farmer
clientele. Since collection points could be dispersed throughout Uganda more easily than a number of other types of livestock
development projects, the centers enabled the ministry to
strengthen its ties to farmers in areas not well served by other
programs and to defend itself against accusations of discrimination in the distribution of its activities among the regions and
ethnic groups of Uganda.[22] These political considerations--
the demonstration of a contribution to the realization of the
goals of Ugandan nationalism and the building of a large and
widely dispersed clientele--overcame the counsels of economic
prudence in rapidly extending the collection network.

The dairy cooperative constituted a problematic factor, interfering with the realization of the political advantages which
the dairy agencies sought to gain through their economic adven-
turism. Cooperatives are likely to prefer high-return sales of
raw milk in local markets, in contradiction to the priority which
the dairy agencies assigned to the movement of milk to the major
urban areas to displace imports. In addition, as relatively autonomous local organizations, mediating between the Dairy Corporation and the farmers, they endangered the dairy agencies' ability to win a full measure of clientele support in instances of
successful marketing operations; according to the Corporation's
secretary: "most of the praise going to the coops I believe
should go to the Corporation."[23] On the other hand, where the
dairy agencies felt that the considerable aid they provided was
being misused, they lacked satisfactory means of intervening to
correct the situation. The formal, legal right to intervene directly in the management of society affairs was in the hands of
the Cooperative Department, which was not subject to the direction of the Ministry of Animal Industry. Both a more patient approach to society foibles and organizational self-interest meant
that the Cooperative Department tended to take the societies'
side in disputes with the dairy agencies. Thus, although the
usefulness of clientele organization was recognized,[24] the
leadership of the dairy agencies came to the conclusion that organization with the degree of autonomy enjoyed by the cooperative
vis à vis the dairy agencies was intolerable, and steps were
taken to reduce the role played by cooperatives in the handling
of milk.

Post-Coup

Amin's coup of 25 January 1971 led to the removal of John
Babiiha from his post as Minister of Animal Industry. Babiiha
had taken a close interest in dairy policy and had placed his
personal imprint upon a variety of major policy decisions, including the position taken by the dairy agencies toward coopera-

tives. With his withdrawal from the scene, the way was opened
for a review and revision of positions on this question.

Changes were urgently needed since the gamble involved in
the expansion of the collection network under the direct administrative control of the Dairy Corporation had clearly proven
unsuccessful. The Corporation's financial situation was disastrous, and its financial and technical operations were in a state
of profound disarray.[25] Rather than enabling the dairy agencies to take credit for successful milk marketing, direct administrative control over the centers had in too many instances
identified the agencies in the farmers' eyes as those responsible
for delays in milk payments and the abusive behavior of the centers' milk handling staff. The agencies' leaders came to realize
that direct administration of local level affairs did not necessarily bring benefit, given their inability to exercise effective
control over their own organizations, and in particular over local level staff. On the other hand, there was an awareness of
the value of society officials as an on-the-spot supervisory
group with a direct personal interest in guaranteeing the efficiency of milk collection and marketing in the rural areas.[26]

The dairy agencies, therefore, took steps to lessen the tension which had come to prevail in their relations with the dairy
cooperatives and the Cooperative Department. Take-overs of society operations were ceased; indeed, in one instance the Dairy
Corporation flatly refused to relieve a society of responsibility
for the operation of its collection center.[27] A favorable attitude was adopted toward the participation of cooperatives in
the operation of new cooling centers being opened in various
parts of Uganda. Contract negotiations were undertaken with a
view to guaranteeing veteran societies security and legally recognized status in all that concerned their relations to the dairy
agencies. The dairy agencies even expressed willingness to consider returning to society control a number of centers which had
been taken out of cooperative hands.[28]

In sum, then, economic-technical considerations, which had
carried considerable weight in the initial decision to involve
cooperatives in the organization of milk marketing, led the dairy
agencies back to the idea of cooperation after the failure of
attempts to do without it. The agencies once again encouraged
farmer organization within cooperatives so as to obtain partners
who would share in the financial and administrative responsibilities which the agencies' own resources alone had proven incapable
of bearing.

Government and Dairy Cooperatives: Conclusion

The attempt to do without dairy cooperatives and their consequent restoration to favor nicely illustrate our general propositions concerning the factors leading to choice of cooperative

rather than direct governmental administrative control of local level activities in the rural agricultural sector. Particularly in new and difficult fields such as milk marketing, the locally based, relatively autonomous cooperative organization serves as a buffer between national political and governmental actors and blame for failures. In addition, the local farmer organization complements governmental resources and skills; and it provides a counterweight to the freedom of action of local level staff, who cannot be adequately supervised and directed by the national authorities. The apparent concession involved in acceptance of cooperatives' relative autonomy is in fact a reasoned response by the upper levels of the bureaucracy and the political leadership to the limits of their ability to convert administrative intervention into politically and economically effective control.

8. The Department of Cooperative Development: Administrative Aspects of Government Intervention

INTRODUCTION

In the relatively limited literature dealing with cooperatives, the administrative instrumentalities of governmental intervention in cooperative affairs receive only marginal consideration. However, administrative characteristics and capabilities
are obviously of great importance in determining the ultimate impact of government policy. More specifically, national political
leadership's ability to realize the developmental goals of efficiency and equality is largely dependent upon the level of administrative performance. The same holds true with regard to
policy-makers' ability to solidify their political standing by
achieving effective control of various aspects of economic and
social activity. Indeed, as we have argued above, policy-makers'
understanding of the reliability and capability of the administrative units at their disposal is one of the factors determining
the degree and form of intervention in the economy. Therefore,
we attempt in this chapter to assist in filling the gap in the
literature by providing an empirical examination of the Ugandan
Department of Cooperative Development.

In the sections which follow, we will first consider the
"ideology" of the Cooperative Department, with a view toward
clarifying the values and priorities guiding Departmental activities as they are understood by the top leadership of the Department.[1] It will be seen that organizational ideology, as expressed by Department heads, is marked by a sober, businesslike
tone and the absence of any note of egalitarian, populistic en-

thusiasm; this mirrors an equally apolitical orientation in the comparable Ghanaian department. We then turn to a discussion of the Departmental leadership's operational orientation--that is, their concept of the role to be played by the Department in relation to the other elements of the cooperative movement in order to make possible the realization of cooperative goals. Having outlined the somewhat contrary demands of the Department's role, we proceed to describe the actual working methods of the Cooperative Department, as observed in the field. We find that the organization of intervention in cooperative activities departs significantly from the model of bureaucratic formalism which is often presented as characterizing the work of government agencies in third world nations. Our analysis shows that deviation from bureaucratic norms has been detrimental to the realization of both efficiency and equality goals. Having presented these rather unusual findings, we close the chapter by suggesting a possible explanation for the failure to maintain bureaucratic procedures in situations where they were clearly appropriate. We contend that departure from bureaucratic norms is not simply a sign of weakness or incapacity, but that it reflects instead the existence of a professional self-concept among Departmental leaders which legitimizes the failure to devote attention to codification of clear policies and detailed, standardized working procedures.

ORGANIZATIONAL IDEOLOGY

The Ugandan heads of the Cooperative Department remained loyal to the staid, business-oriented concept of cooperation which was held by their British predecessors. The Department of Cooperative Development was founded in 1946, after the enactment of the Cooperative Societies Ordinance, so as to provide the colonial government with the ability to build an orderly, government-regulated, and--as the British saw it--apolitical alternative to the spontaneous, mushroom growth of cooperatives which had combined economic demands with African nationalist political values and aspirations. The colonial government, in the language of the Cooperative Department's annual reports, recognized the need to encourage African participation in the marketing and processing of their own crops. By itself working through cooperatives, the government hoped to undercut the nationalist monopoly on mass organization by means of economic appeals and services. Also, government powers of intervention in cooperatives helped guarantee against the subversion of Africanization efforts by interested Asian parties, operating through companies formed by African front-men.

For the British, then, cooperation was a paternalistically

supervised movement to promote farmer interests, as against those
of commercial and processing middlemen. It was the colonial government's professional, apolitical response to its perception of
concrete economic needs of a broader sector of the population and
its own desire for control and stability.   The most value-laden
expression to appear in the annual reports of the British heads
of the Department is "mutual help."[2]

Despite the populist and socialist rhetoric which began to
become prevalent within politial leadership circles in independent Uganda, the Ugandan heads of the Department continued to
present their organization and its activities in terms virtually
free of any reference to the differences in wealth and status
within the agricultural community and possible conflicts between
those at different levels in the stratification system. A catch
phrase appears occasionally, but the concept of conflict between
rich and poor which lies behind such phrases is never developed;
rather, the businesslike tones of the reports of the colonial era
continue to predominate.   Even a project like group farming, with
its obvious ideological significance, continued to be described
by the bureaucracy in purely technical and commercial terms,
without reference to broader social and value implications of the
undertaking.[3]

Amin's seizure of power, of course, cut short the elaboration of a populist-socialist ideology in Uganda, replacing the
UPC political leadership with a personalistic tyranny lacking
such intellectual aspirations. Whatever pressures the leadership
of the Cooperative Department may have felt in the late Obote
period toward explicit analysis of the socioeconomic structure
of rural society and the role of cooperation within it were removed by the change of regime. Examining what would seem to be
the most comprehensive self-description of the Department ever
presented by its senior officials--an essay dated 28 September
1973, and entitled "The Duties and Functions of the Department
of Co-operative Development"--one is struck primarily by the degree of continuity in the conceptualization of cooperation as a
businesslike combination of individuals for the efficient promotion of shared economic interests.[4]

According to the author, "In the modern sense, a cooperative
society is a business organisation formed by people for the purpose of providing goods or services or both for themselves."[5]
The cooperative, it is noted, has certain distinguishing characteristics which mark it off from other business organizations,
most particularly open membership and democratic control; but
the description of the cooperative as a business indicates leadership's conceptualization. Cooperation is a tool available to
those willing and ready to work together to help themselves and
each other; the idea of cooperation is not in any sense an element in a critical analysis of society and economy, informed by

abstract social values, wherein cooperation appears as an instrument which by its very nature is adapted to serving the interests of the oppressed in their struggle for social justice.

The apolitical nature of the approach stands out strikingly
in the paragraph in which the author attempts, more than in any
other section of his essay, to depict the values realized through
cooperation and the broader social significance of cooperative
organization. After summarizing in monetary terms the extent of
cooperative assets and the scope of cooperative economic activity, he states:

> The Co-operative movement has not only expanded in busi-
> ness but has taken development to rural areas. Today,
> there is nearly in every village a store [that is, a co-
> operative warehouse] where other business such as shops,
> schools, community centres have sprung around. Through
> co-operatives Government has managed to provide credit
> and other services to farmers without security. The co-
> operatives have provided some education to its members
> in management and honesty and trained some members to
> become public spirited. It has aroused the interest of
> ordinary man to work hard and also realise that he can
> also play a role in the economic development of his coun-
> try. It has brought about orderly marketing of produce,
> raised the standard of quality of produce marketed. It
> has taught people even in the remote area to be self-
> reliant and democratic in their operation.[6]

Cooperation, then, offers certain advantages to the average
farmer; it helps him to be better integrated in national development efforts. However, note what is lacking here: we find no depiction of the relations between average farmers and the wealthy,
or the relations between both groups and the public authorities
which have become so important in dispensing benefits and opportunities to the farming community. Neither does one detect any
sense of cooperation as a means of equalizing the situation of
the advantaged and the disadvantaged.

Operational Orientation

Ambiguity marks the Cooperative Department concept of its
role in the realization of cooperation. On the one hand, success
as measured by commonly accepted business criteria is essential
from the viewpoint of both the societies and unions and the Department itself. Bearing responsibility for the commercial success of cooperatives, the Department has naturally sought the
means necessary to discharge its responsibility via control over
cooperatives' decisions and operations. During the period between independence and the enactment of the Cooperative Societies
Act of 1970, the Department complained of the need for revision

of the 1963 Act so as to place more decisive powers in governmental hands.[7]   This trend in Departmental thinking also finds
expression in "Duties and Functions": "The Government and through
the Co-operative Department has and it is still doing all within
its power to assist, guide and <u>control</u> the affairs of the Cooperative Movement [emphasis added]."[8]

  However, the formal definition of cooperation includes the
element of democratic participation and membership control of cooperative operations. Success on the commercial front resulting
from the appointment of Department staff as supervising managers
constitutes an admission of failure in achieving the goal of edu-
cating the mass of cooperative members to a degree of informed
activism capable of guaranteeing responsible and capable performance by cooperatives' elected leadership and hired professional
staff. However, it is quite simply beyond the resources of the
Department to provide direct operational guidance for the entire
complex of unions and local societies. Thus, in parallel--and
in contradiction to--the drive for increased formal powers of intervention, Department leaders depict the Department as adviser
and critic and auditor, lacking both power and responsibility to
undertake direct management of cooperative affairs. Responsibility lies with the elected leadership, in the first instance, and
ultimately with the membership:

> In Uganda, the co-operative societies are actively
> supported by the Government and this sometimes is mis-
> taken by the majority of population to regard them as
> part of the Government machinery, but strictly the co-
> operative societies are corporate bodies which can sue
> and be sued. The societies select their own leaders and
> they have got a right to remove them whenever members are
> not satisfied with their leadership.
>
> The Co-operative Department is not part of the Move-
> ment. It is a Government Department which is put there
> by the Government to check and supervise the Movement.
> Many people appear erroneously to regard the department
> as part of the co-operative movement and whenever a mis-
> take is done by the movement it is put against the Co-op-
> erative Department. But one thing that should be under-
> stood is that the Co-operative Department like any other
> Government Departments . . . is just an adviser to the
> public. The public may decide to ignore the advice.[9]

In sum, then, the heads of the Department of Cooperative Development saw their organization as called upon to guarantee honesty
and efficiency in cooperative operations, through the exercise
of maximum practical control over cooperative affairs--without
being forced to take drastic steps of direct administrative intervention, such as the transfer of the powers of elective officials to an appointed supervising manager.

WORKING METHODS

Our field observations suggest that the Department's working methods in practice are marked by a striking lack of attention to bureaucratic routines. Hierarchical supervisors devote relatively little attention to the codification of detailed operating procedures and uniform decision rules which will oblige their subordinates. We will develop this point through a number of examples.

The repetitive cycles of the financial year's bookkeeping and accountancy lie at the heart of the work of the Department's field staff. The Departmental leadership's choice of accounting systems and design of basic reporting documents--annual accounts and balance sheet, requests for authorization of distribution of surplus, requests for authorization to assume financial liabilities--fix certain operating routines which do have a great impact on the activities of Department representatives in their contacts with the societies and the unions.

However, evidence available concerning dairy cooperatives indicates that full definitions of accounting procedures are not formulated. Thus, for example, in the accounts of some dairy societies a variety of expenditures appear in the trading account (the first major heading of the standard annual account) and thus are deducted from the societies' gross returns from the sale of produce in the calculation of the societies' gross surplus. In other cases, such outlays appear in the expenditures column of the income and expenditure account, which follows the trading account. Thus, rather than being deducted in the calculation of gross surplus, they are deducted from the gross surplus in order to arrive at the net surplus. These variations make analysis of the situation of groups of societies--their problems and possible solutions--impossible without rearrangement of the accounts and recalculation of the various indicators, which should have a standard definition. Indeed, the varying treatment of certain types of expenditures may make the accounts of a single society incomparable from year to year without refiguring. Yet these deviations from uniform format pass without comment; such accounts may be found in district and national headquarters' files without record of protest on the part of either general supervisory personnel or officers with specialized responsibility for audit or livestock.

More significant than this point of inconsistency in society accounts is the typical lack of detailed recording of milk sales and purchases. In some instances, the trading account provides information concerning the quantities of milk purchased and sold at various prices. This is, of course, vital information to any attempt to gain an informed understanding of the market situation affecting the societies' purchases and sales of milk and the efforts devoted by society leadership to the development of sources

of supply and favorable sales outlets. Nevertheless, in most annual accounts of dairy societies, quantity and price information
are not recorded in the trading account.[10]
    Further, our East Buganda dairy cooperative material sug-
gests a more basic problem: though accounting cycles could con-
ceivably have served as a foundation for Departmental routines,
top-ranking officials devoted little effort to building a strong
structure of organizational procedures and output standards upon
the professional element of accountancy. Thus, for example, firm
guidelines were not established concerning required minimum fre-
quency of visits to societies and the audit and inspection activities which form the basic element of staff-society contacts.
In parallel with the lack of such guidelines, systematic review
of files by supervisors to verify fulfillment of audit and supervision requirements was also not undertaken. Rather, work sched-
uling was left to the energy and initiative of local staff; and
crisis, rather than periodic review, tended to draw supervisory
staff's attention to failings in the maintenance of regular supervisory contact by field staff.
    Thus, for example, the District Co-operative Officer/East
Buganda writes to the Co-operative Assistant responsible (through
the county Assistant Co-operative Officer) for the Bugerere Balunzi Society, which was at that time plunging into a deep financial crisis:

> Records in my office show that the above society was
> last visited, for the purpose of interim audit and gen-
> eral inspection, during the month of July 1972 [eight
> months previous]. . . .
>   . . . The fact that you were allocated four-five so-
> cieties directs you to pay much attention to those soci-
> eties. These societies should be visited at least once
> a month or more as in the case of Bugerere Balunzi Dairy
> Co-operative Society, Ltd.[11]

Rather than being fixed in advance, as a general rule, the standard is fixed here after the fact, and as a more or less personal
admonition to a particular officer.
    Another example drawn from our East Buganda field work to
demonstrate the lack of authoritative and formally established
standards lies in the sphere of registration of cooperatives.
One of the crucial issues in the Co-operative Department's treatment of dairy cooperatives was the question of criteria for the
formal, legal registration of societies dealing in milk. Government recognition of a society was necessary to establish exclusive control over milk marketing. Moreover, after the 1971 coup,
the dairy agencies decided that only registered cooperatives
would be considered for allocation of milk coolers; thus, official recognition became an essential step toward the acquisition

of this vital item of equipment.[12] Finally, and most basi-
cally, registration was necessary so that the cooperative, as
such, could function as a legal personality and enter into commercial and financial agreements. Without registration, even the
opening of a bank account in the society's name was difficult,
and the cooperative's operations had to be carried out by officers, acting formally as individuals, but in fact acting in the
name of the society.

Who, then, had the right to be registered as a cooperative
society? The law fixes a number of requirements which every society must meet:

> 2. Subject to the provisions of this Act, a society
> which has for its object the promotion of the economic
> interests of its members in accordance with co-operative
> principles, and which in the opinion of the Registrar is
> capable of promoting those interests, may be registered
> under this Act. . . .
>
> 3. No society shall be registered under this Act un-
> less,
>
> (a) it consists of at least ten persons all of whom
> are qualified for membership of the society . . . .[13]

Interpretation of the broad principles of Paragraph 2 has
fallen to the Cooperative Department. It is striking that despite the most basic nature of the question, the Departmental
leadership has not promulgated unambiguous, detailed standards
and operating procedures for their application. Instead, recommendations for the registration of societies are left to the professional judgment of the District Cooperative Officer, who relies in turn upon the evidence and evaluations presented to him
by his field staff. These recommendations usually touch primarily upon the potential economic viability of the societies, and
make little mention of the question of faithfulness to the spirit
and values of cooperation.

This finding concerning the lack of clearly defined criteria
for registration and Departmentally fixed evaluation procedures
may be illustrated by examining the case of the registration of
the Kimwanyi Dairy Cooperative Society. The Kimwanyi Society was
registered in May 1973, despite a number of objections raised by
the Assistant Cooperative Officer/Livestock for East Buganda District, the Animal Husbandry Officer/Dairy in charge of Bulemezi
and Buruli Counties, and the neighboring Semuto Livestock Cooperative Society, which had been registered in July 1972. The main
points raised against the registration were as follows:

> 1. The society's base of operations, Nakaseke trading
> center and the surrounding Gombolola [subcounty] Saba-
> gabo, fell within the area of operations which had been

approved for the Semuto society upon its registration. A considerable group of Semuto's active members were concentrated in one of the Sabagabo parishes.

2. Nakaseke is situated less than twenty miles from milk coolers located at Semuto and Katikamu. Thus, according to dairy agency guidelines fixed in the post-coup period, Nakaseke could not be considered for allocation of a bulk milk cooler, as required for large scale milk collection operations.

3. The society's main business, the trade in milk, was conducted on an extremely small scale. Only two or three of the forty-three fully or partially paid-up members of the society delivered milk on a regular basis. None of the officers supplied milk to the society. Sabagabo, the society's proposed area of operations, seemed unlikely to provide sufficient quantities of milk for viable commercial operations.

4. Suspicions were raised concerning the genuineness of a number of the registered memberships. It was suggested that to make a better impression, the number of members had been inflated by registering several individuals from a single family as members, though they all represented the same farm.[14]

In sum, the Society's current performance and future prospects of commercial success were unencouraging. Aspects of the Society's operation seemed to depart from the principles of cooperation and, indeed, with regard to point 4 (membership) from basic standards of frankness and honesty.

In the last analysis, we must point to the high status and influence of Society officers and leading members as crucial in bringing about the registration of the Society despite the strongly negative evaluations of the field staff directly responsible for dairy cooperatives and the initial inclinations of the DCO.[15] However, at the same time, it must be noted that the lack of clear standards defining the suitability of a society for registration weakened the hand of those inclined to deny Kimwanyi the official recognition which it sought. Points seemingly disqualifying the society were noted, but at no stage could those against the proposal point to an unambiguous Departmental policy document, defining registration standards, in order to shield and justify their judgment against the internal and external pressures in favor of registration.

The question stands out most clearly with regard to the issue of minimum conformity to the principles of cooperation. However difficult a professional problem the prediction of economic viability may be, adherence to certain basic principles of cooperation could be clearly operationalized. Thus, for example, consider the question of inactive (if not actually "dummy") mem-

berships, on the one hand, and, on the other hand, a considerable
traffic in the goods of nonmembers. "Duties and Functions" does
state that "Non-users should not be members of a society," but
cannot cite an official prohibition.[16] Proscription of trade
in nonmember produce lies buried away in the standard bylaws in
which the specific details of each society are inserted for registration. However, examination of conformity to this rule is
not an obligatory part of the registration process, and no minimum standard, in terms of percentage of produce supplied by members, is fixed to guide staff evaluating particular societies
being considered for registration.[17] In the case before us,
then, district staff might make known their feeling that: "if
about 3 money magnets [sic] form a society under the guise of
another 50 poor members, this will not help the poor members very
much. If this is the case these individuals could just as well
form a partnership."[18] However, field workers could not state
outright that the registration of the society was forbidden according to Departmental criteria, and thus the grant of official
recognition to a "group [which] is more or less a company dis-
guised as a society" was facilitated.

Implications of Non-Bureaucratic Working Methods
for the Realization of Development Goals

It is common to speak of the existence of a basic dilemma
in development administration: the unsuitability of elements of
the bureaucratic model of administration to the conditions of
development.[19] Our review of the functioning of the Ugandan
Department of Cooperative Development suggests that this widely
accepted image of the development administration situation is not
as universally applicable as generally assumed.[20] In the case
we have examined the generalization is incorrect, both empirically and normatively.

Empirically, the Cooperative Department is not a rigidly
bureaucratic organization in which stifling hierarchical control
is exercised over operational staff through a detailed set of
written regulations, together with strict reporting procedures.
The image of the hyper-bureaucratization of governmental administration in the third world nations has been shaped largely by the
Indian and, more generally, the South and Southeast Asian experience.[21] Here the influence of the generalist administrator,
serving in the ranks of an elite administrative corps, is reported to have left its imprint upon the functioning of the entire civil service. By contrast, in Uganda and, one may suggest,
most of Africa, there is no elite civil service corps with overwhelming prestige and influence permeating all levels and departments of the governmental apparatus. Nonspecialists are to be
found in positions in the office of the minister in the ministries, and the District Commissioners have important functions

in their respective districts. However, effective direction of
the functional departments at the national and field levels is
not provided by these officers.[22] Indeed, one may say without
fear of exaggeration that in shaping the operational outlook and
style of the substantive departments of government, the generalist administrator's influence is hardly felt.

Normatively, we have shown that in many instances departure
from the bureaucratic model, rather than bringing a positive
flexibility, simply leaves less well-trained and experienced
staff in the field with insufficient guidance and supervision in
the conduct of their duties. Moreover, since the lower ranking
staff are not subject to detailed instruction, supervisory staff
at all levels lack clear standards to direct their oversight activities.[23] Lack of uniformity in operations and recording,
resulting in the first instance from lack of Departmentally established procedural and substantive norms, also means that accumulation of comparable data from the field for analysis is
impeded.

It may be argued, then, that a point made by Lowi with regard to American national politics holds as well with regard to
the internal development of administrative units in the developing countries. Clean definition of goals and obligatory measures
for their realization makes for good administration; it provides
the standards for oversight of current operations and facilitates
learning concerning the field of activity in question, thus making possible the amendment and improvement of policy.[24] Development is indeed a learning process; but, as much as fluidity
and uncertainty demand a free and flexible approach in certain
truly unexpected and complex situations, they demand as well a
constant effort to codify the experience which has accumulated
in organizational regulations, so as to provide a base-line
against which new situations may be evaluated. Great effort must
be devoted to giving reactions to situations lying outside the
range of organizational experience something of the systematic
nature of experimentation, from which accumulation of knowledge
is possible.[25] In the Ugandan case, the effort to structure
and codify has received short shrift.

The application of Lowi's insights may be carried one step
further: in the districts of Uganda, as in the regulatory commissions of the United States, lack of clearly defined standards
opens the way for political pressures. An ambiguous set of directives leaves those expected to prevent abuse of policy without
a clear organizational imperative to serve as guide, justification, and presumptive guarantee of backing from superiors when
taking steps which arouse resistance. Where standards are lacking, administrators and elements in the environment reach accom-
modation through bargaining. In our case, lacking a clear criterion demanding and justifying outright rejection of the Kimwanyi application, the District Cooperative Officer and others

negotiated with the Kimwanyi and Semuto Societies, and, as in Lowi's American examples, the strongest actor most directly involved in the questions under administrative care succeeded in striking the bargain he wished.

The normative implications of our analysis of what we have termed premature professionalization are thus clear. For the sake of efficiency and, even more emphatically, for the sake of equality objectives, it may well be necessary to impose a more bureaucratically oriented approach to implementation, involving detailed definition of tasks and close oversight of performance. We have seen above that even the limited, nonradical concept of cooperation held by the Cooperative Department leadership could not be enforced in field work without an effective operationalization of its principles. Implementation of a more radical approach, had the UPC government persisted and attempted to formulate one, would certainly have been doomed to failure without a change in the Cooperative Department's operational style.

Why Non-Bureaucratic Working Methods?

How then shall we understand the very considerable departure of Cooperative Department working methods from bureaucratic patterns? We contend that the lack of effort devoted to the elaboration of decision-making criteria and the codification of operating procedures cannot be explained solely on the basis of such factors as limited numbers of highly qualified staff and the great burden of work imposed upon the Department's officers. Our findings have shown the absence of clearly defined official instructions with regard to fundamental questions of workload/output norms, accounting procedures, and the qualifications for registration as a cooperative society. The elaboration of norms and procedures for less central issues might be pushed aside by the press of day-to-day responsibilities, but the failure to deal with truly basic subjects demands a more complex explanation.

We would suggest that the heads of the Cooperative Department emphasize the role of "professionalism" rather than bureaucratic hierarchy in Departmental operations.[26] The Cooperative Department, like a number of other Ugandan government departments, is headed by officials whose background is that of subject-matter specialists, rather than administrators.[27] They are among the first African recruits selected to receive full professional training. Their standing as African pioneers in the professions upon which Department functions are based allowed them rapid advancement to top positions of Departmental leadership. It is, therefore, not surprising that these officials' concept of the Department's and their own role continues to be shaped by a primarily professional, rather than an administrative, outlook.

Thus, the Acting Registrar/Commissioner for Cooperative De-

velopment writes in "Duties and Functions": "To put it in a nut-
shell, the duties and functions of the Registrar and his Senior
Staff demands on them to be fully professional accountants, managers, lawyers, financiers, educators, economists and planners
and mature administrators."[28]   The first seven professional
roles are described in some detail in terms of the services which
the senior staffer must be able to perform on behalf of cooperative organizations, officials, and members. By contrast to his
extended treatment of professional duties, the author fails to
analyze the term "mature administrator," as it relates to the internal organizational responsibility of the Department's leaders
--their responsibility for the development and guidance of the
Department itself. For the most part, a similar imbalance appears in the leadership's depiction of Departmental activities
presented in the annual reports.

"Education"--that is, professional training--and the provision of adequate numbers of highly qualified staff are the only
organizational issues to receive a consistent emphasis in the
annual reports. Here, too, the professional approach is evident.
Organization is adapted to professional imperatives: training
guarantees the Departmental staff member's ability to function
independently in keeping with universally recognized principles
of cooperative practice. The Department seeks to recruit and
train able personnel, to obtain the establishment rankings and
budget needed to hold qualified staffers within its ranks--and
relies upon the professional competence of staff to guide its
work, rather than Departmental hierarchy and detailed direction
imposed from above. Emphasis upon professional training is,
thus, an alternative to the devotion of time, energy, and priority to the internal condition of the Department as an organization.

The operations of the Department, therefore, are marked by
what might be termed "premature professionalism," rather than the
frequently decried hyper-bureaucratization. Departmental leadership lacks the inclination and the habit of formulating detailed
sets of rules to guide the actions of Department staff. The
failure to elaborate a corpus of organizational regulations
obliging field workers is legitimized by professional values.
These values attribute central importance to the informed professional judgment of the staff member directly responsible for the
treatment of each particular situation under the Department's
care. Unfortunately, for reasons we have discussed above, such
an approach has negative implications for the realization of both
equality and efficiency objectives.

9. The Political Economy of Ghana: An Overview

COCOA AND THE COLONIAL STATE IN GHANA

If you want to send your children to school, it is cocoa,
If you want to build your house, it is cocoa,
If you want to marry, it is cocoa,
If you want to buy cloth, it is cocoa,
If you want to buy a lorry, it is cocoa,
Whatever you want to do in this world,
It is with cocoa money that you do it.

-- Ashanti ditty[1]

We now direct our attention to the political economy of cooperation in Ghana. Again, to indicate the context within which
the cooperative drama evolved, we need first to provide an overview of the political and economic background which shaped Ghanaian cooperation. The cooperatives we examine functioned in the
cocoa sector; accordingly, this introductory section will concentrate upon political and economic aspects of this crop, which
has dominated the Ghanaian economy since early in the twentieth
century.

British presence in coastal slaving entrepôt outposts in
Ghana dates from 1650. A fluctuating zone of influence emerged
in the near hinterland, with Fanti middlemen operating increasingly under the shadow of British patronage. However, no claim
of sovereignty was made until 1874. The coastal areas alone
formed the embryo of the Gold Coast; beyond lay the domain of the
powerful Ashanti confederacy, a commercial-military state which

took shape about 1700.[2] A series of wars set the British-sponsored forces and the Ashanti against one another during the nineteenth century; in 1895, the British mounted a decisive assault on the Ashanti capital, seized and exiled the King, and incorporated the Ashanti domain. The way then lay open for the extension of the colony northward to the 11th parallel by 1899. After World War I, the western portion of former German Togoland was added as a League of Nations mandate, completing the territorial realm which is Ghana today.

Initially, the economic resource which most attracted British attention was gold; bullion had been exported to Europe for many centuries across the Sahara. However, the most portentous economic change was the introduction of cocoa, almost entirely through African initiative. A Ghanaian entrepreneur returning from a period of labor in Fernando Po in 1879 brought some pods with him, and by the mid-1880s cocoa beans were exported to Europe. While British firms soon established their control over the export sales of the cocoa, and the missions played some role, the actual spread of cultivation was a remarkable saga of African ingenuity and adaptability, accomplished in the face of lack of support and even hostility from the colonial administration until after World War I.[3] Table 9.1 chronicles the speed of this success.

Two somewhat different cocoa development patterns emerged. About half the cocoa zone lay in Ashanti, where production was

Table 9.1

COCOA EXPORTS, FIVE-YEAR AVERAGES
(tons)

Years	Five-Year Average
1891-1895	5
1896-1900	230
1901-1905	3,172
1906-1910	14,784
1911-1915	51,819
1916-1920	106,072
1921-1925	186,329
1926 1930	218,895

Source: F.M. Bourret, Ghana: The Road to Independence (London: Oxford University Press, 1960), p. 23.

undertaken by local farmers, initially with family labor and some rental of land. Most of the remainder is located in the Akwapim hills and adjacent areas, sparsely inhabited forest land put into production by Ga, Krobo, and other migrants, as well as local residents. The migrants formed "companies" to negotiate land rights from local chiefs, but farmed individually.[4]

Cocoa soon became the crucial base for state revenue. A cocoa export tax was first levied in 1916, ostensibly as a war measure. The fiscal dependence of the state on cocoa quickly became permanent, with the tax ranging from 12 to 28 percent of the export value.[5]

The dominance of cocoa in the economy became institutionalized; by the interwar period, it generally accounted for over 70 percent of the value of exports. After the extraordinary pace of early development, expansion began to slow. Output averaged 218,895 tons from 1926 to 1930, and 236,088 from 1931 to 1934. In 1936, a record of 311,151 tons was set, which was not reached again until 1959. Much of the land best suited to cocoa had by now been put in cultivation. In the 1930s, swollen shoot disease began to affect the crop. No doubt most important of all, the price was unfavorable through much of the period. After a short-lived boom in 1919–20, when the price briefly soared to 120 pounds per ton, the price dropped; in 1929, it was 50 pounds per ton, then fell in 1930 to 20 pounds.

Though European export firms—especially the United Africa Company—controlled export trade, they dealt through African buying agents who were often also cocoa farmers. Cocoa, unlike cotton and coffee, requires no capital infrastructure for processing. The harvested pods are broken open on the farm, and the beans fermented and dried. Storage, buying stations, transport, and financing must be supplied by brokers and buying agents; these commercial skills were easily available in the cocoa regions of southern Ghana, and the interface between the peasant farmer and the export trade was the African buyer, even though he was tied to the European firm. In the 1930s, there were as many as 40,000 African brokers and buyers.[6]

European capital, led by the ubiquitous United Africa Company, did seek to forcibly enlarge its share of the proceeds in cocoa exports, by forming a secret buying combine in 1937 whereby the 14 firms dealing in Ghana cocoa would control prices paid to the African brokers. When word of this price-rigging scheme leaked out in late 1937, cocoa farmers, organized by chiefs and nascent farmer organizations, organized a remarkable boycott. Not only did they succeed in almost totally halting all cocoa sales, but the retail outlets of the firms in the combine were also shunned. The producers were able to maintain their strike for seven months, at the cost of great privation and sacrifice. A parliamentary commission was finally dispatched from London to mediate the conflict; while they did not give total support to

the growers, the buying ring was broken, and the commission concluded that Africans had been almost 100 percent behind the boycott movement.[7]

In the cocoa areas, an inter-connected socioeconomic elite
of chiefs, cocoa merchants, and some larger farmers emerged, with
some overlap of roles.[8] The cocoa economy, however, did not
give rise to latifundia agriculture. The great bulk of production occurred on roughly 300,000 farms, averaging 6-7 acres, with
about a ton of output. A new substratum began to be visible, as
a number of cocoa farmers began hiring labor, which was mostly
supplied by migrants from the impoverished zones to the north,
especially neighboring Upper Volta. Labor was hired on a wage
basis for the heavy work of opening new farms. Larger farmers
with a number of plots often left established groves in the cus-
tody of abusa caretakers, who received one-third of the crop in
return. However, most farms were predominantly family units,
with some seasonal labor at bottleneck periods of the year.[9]

THE NATIONALIST ERA: TERMINAL COLONIALISM IN GHANA

The postwar years saw profound changes. In Ghana, the colonial state, in its final phase, vastly enlarged its scope of
operation. In 1946-47, public expenditures were still only
6,630,000 pounds; in the previous 35 years, they had increased
only twofold. During the final decade of British rule, they multiplied by ten, a measure of the wholly new engagement of the
state in education, health, roads, and other public amenities,
and development services.[10]

Soon after the war, the burgeoning colonial state came under
African assault. Higher consumer prices, unreasonably low prices
for cocoa, the ravages of the cocoa swollen shoot disease, and
the discontents of returned servicemen created an underlying malaise, while the new anti-colonial spirit in the world and the
Asian surge to independence generated a sense of anticipation of
impending changes. The British initiated a series of reforms in
the immediate aftermath of the war, with a view toward gradual
incorporation of African participation in the organs of the colonial state.

The formula for transfer of sovereignty, however, was not
destined to follow the genteel pattern blueprinted by the colonial administration. New social forces were mobilized, and, in
three decisive years from 1948 to 1951, the colonial state was
in effect conquered by radical, populist elements, whose elan was
symbolized by Kwame Nkrumah and the Convention Peoples Party
(CPP). Nkrumah, after a dozen years of study and work in America
and Britain, returned to his homeland at the end of 1947, as organizing secretary for the United Gold Coast Convention Party,
representing a professional and chiefly elite. Two months later,

the long tranquility of colonial order was shattered, when two weeks of bitter rioting broke out triggered by the discontents of ex-servicemen (who numbered 63,000).

Events then moved swiftly. In June 1949, Nkrumah with a number of younger leaders broke with the UGCC, to launch the CPP. A new, aggressive, agitational style of politics was initiated. The activist style and populist message found a ready audience among newly politicized groups: lorry drivers, market women, government clerks, ordinary cocoa farmers, and perhaps above all the swelling army of primary school leavers. These social categories could not easily accept the coastal intelligentsia or the chiefs as automatic heirs to the keys of the kingdom.

The first natonal elections in 1951 were a decisive triumph for the CPP, which won 29 of the 33 rural electoral colleges, and all 5 municipal seats. The colonial administration accepted the results with good grace, and Nkrumah--then jailed on sedition charges--left prison to become Leader of Government Business. The CPP had scaled the heights, and taken the state by storm in bloodless combat.

However, troubles lay ahead. As the second national election approached in 1954, various focal points of communal discontent became visible. Most important were the strong anti-CPP sentiments which emerged in the powerful Ashanti community. Grievance over cocoa prices was the precipitant, but the dispute soon spilled over into prerogatives of the chiefly hierarchy, and finally the claim that Ashanti, as a historical state in its own right, required autonomous status in the new Ghana. Elements of the old alliance between coastal professionals and hinterland chiefs, eclipsed in 1951, stitched together these disparate elements of discontent into a new opposition movement, ultimately labeled the United Party.

The CPP triumph was less sweeping in 1954, though still decisive; with 55 percent of the votes, the party took 72 of 104 seats. The scale of violence and dissension was great enough, particularly in Ashanti, for the British to require that the CPP demonstrate its mandate one more time in 1956 before independence was granted. The results were a confirmation of 1954; with 71 of 104 seats, and 57 percent of votes cast, the CPP was clearly the dominant formation.

The economic underpinning of the political kingdom won by Nkrumah and the CPP remained overwhelmingly cocoa. Wartime exigencies led to a major restructuring of cocoa marketing, with wholly unperceived consequences in introducing state-directed oligopoly. The British government, when war broke out, decided to buy the entire cocoa crop, with a minimum price to be announced at the start of each season. A cocoa control board, germ of the future Cocoa Marketing Board, was created in London. The extant trading houses were licensed to carry out the actual purchase, grading, and transport operations, with a guaranteed profit through a cost-plus fee schedule. Ghanaians in the cocoa sector were indignant at the excessive advantages afforded European mercantile interests, who were assigned 88.2 percent of the crop, while African cocoa merchants were limited to 11.8 percent, and new entrants were barred. The cocoa control board was chaired by chocolate baron John Cadbury, confirming Ghanaian suspicions that the new arrangements were designed of, by, and for the British trading houses.[11]

In 1947, this wartime emergency restructuring of the cocoa industry acquired a permanent cast, when the Cocoa Marketing Board was established with powers to grade, export, and set producer prices. This latter power swiftly became an engine of surplus extraction from the farming sector. The CMB adduced an all-purpose, twin argument that price policy should be guided by the benign intent of protecting the farmers from the excessive fluctuations of the world market by accumulating surpluses in fat years, and subsidizing the price in lean; at the same time, inflation should be stymied by the therapy of avoiding excessive payments to farmers when the world price was high. By prewar standards, which continued to serve as subconscious reference points, the world prices in the 1947-57 decade ran from good to extraordinary. The real effect of the "stabilization" policy, accordingly, was to withhold from producers huge amounts at high price periods, and only in 1955-56 to provide a price subsidy. By 1951, the CMB had accumulated 76,000,000 pounds in reserves, which soon came to be viewed as a major source of government capital expenditure. From 1951, the government moved to directly tap the cocoa surplus by sharply raising the export duty; the yield of this tax rose to 38,000,000 pounds in the peak 1954-55 year.[12] While the government take of the export value was only 5 percent before World War II, in 1954-55 the state extracted over 60 percent of the total.[13] The reserves generated by the CMB were mostly held in long-term British government securities at very low interest rates (0.5 percent before 1950, 2 to 4 percent thereafter). These and similar funds from other colonies played no small part in financing postwar recovery in the mother country.[14]

Cocoa prices became a subject of bitter political controversy through the period. (See Table 9.2 for a review of production and price data.) There was a significant increase over the prewar norm; in 1937-38, the farmer received 7 shillings, 6 pence per 60-pound headload, a figure which rose to 14/6 in 1945-46, 27/6 in 1947-48, and 65 shillings in 1948-49. Thereafter the peasant price was held within a narrow range, while the world market price continued to improve, rising from 139 pounds per ton in 1948-49 to a peak of 450 pounds per ton in 1954. Once the CPP was in power from 1951, it quickly perceived its own interest in nourishing the revenues of the state. Thus the price was raised in 1951, but lowered the next year. Nkrumah himself solemnly re-

Table 9.2

COCOA PRODUCTION AND PRICES, 1946-57

Year	Production (tons)	Average Export Price	Producer Price (pounds/ton)	Producer Price as % of Export Price
1946/47	197,000		50	
1947/48	208,000	201	75	37.3
1948/49	278,000	137	121	88.3
1949/50	248,000	178	84	47.2
1950/51	262,223	269	131	48.7
1951/52	210,663	245	149	60.8
1952/53	246,982	231	131	56.7
1953/54	210,693	358	134	37.4
1954/55	220,198	353	134	38.0
1955/56	228,789	222	149	67.1
1956/57	264,375	189	149	78.8

Source: Bob Fitch and Mary Oppenheimer, Ghana: The End of an
Illusion (New York: Monthly Review Press, 1966), p. 41;
F.M. Bourret, Ghana: The Road to Independence (London:
Oxford University Press, 1960), p. 232.

cited the classic arguments against allowing producers to reap
the dividends of high world prices:

> it became clear to me [in 1954] that further steps were
> necessary to control the price paid locally to the cocoa
> farmers, otherwise we would shortly be faced with infla-
> tion. . . .
> Cocoa . . . belongs to the country and it affects
> everyone, so we have to think of the general public as
> well as the cocoa farmers. I considered that it was not
> in the best interests of the country to be subjected to
> considerable or frequent fluctuations in the local price
> paid to the farmers and that the present price . . . was
> both fair and reasonable and offered an incentive to in-
> creased production. . . .
>   . . . the funds that accrued to the Government would
> be used on expanding the economy of the country as a
> whole with special emphasis on agriculture. It was im-
> portant to ensure that future generations should be
> spared the dangers of having to rely on cocoa as their

only mainstay, as we had had to do. Also, by using cocoa
funds for development and for providing amenities, it
would be possible to improve the general standard of living in this country as a whole at an early date.[15]

Pursuant to these sentiments, the CPP enacted a law in August 1954, fixing the producer price at 7 shillings for the next
four years. Farmers in the cocoa constituencies had fresh in
their minds the campaign pledges of CPP candidate to raise the
price to 100 shillings. The CMB inflamed the situation with a
press release advancing another shopworn colonial argument for
low prices, suggesting that higher returns would lead farmers to
work less. Reactions were particularly violent in Ashanti, where
half the cocoa was produced; a price of 150 shillings was demanded, and a spiral of agitation began which decanted into the
turbulent National Liberation Movement (NLM), unruly advocate of
Ashanti autonomy.

Finally, it was in this period that the swollen shoot disease became a major issue. First detected in the 1930s, this
virus, spread by a mealie bug, causes gradual decline in the vitality of a tree, and eventually its death. The agricultural
service could find no other means of control than destruction of
the infected trees, plus the healthy ones nearby. The government
after the war undertook a compulsory destruction campaign, which
quickly encountered great hostility. By the end of 1947, an estimated 45 million trees were infected (with an estimated spread
of 15 million per year), out of a national total of 400 million;
2.5 million trees had been cut down. The farmers were uncon-
vinced of the necessity of the government remedy, which removed
trees still bearing healthy pods, and left them to support the
cost of five years' delay before new plantings were ready to produce. The farmer counter-arguments maintained:

> the disease was nothing new but would disappear if left
> alone for a few years; it was only necessary to cut out
> the infected parts and not the whole tree, for how could
> a diseased tree bear good fruit? What was the point of
> cutting out the cocoa trees if the forest trees were left
> to harbour the mealy-bug which was said to carry the vi-
> rus? The government was either wicked or foolish because
> labourers sent into the farms cut down healthy as well
> as diseased trees unless bribed not to. Was there, in
> fact, a secret motive behind government action? Did it
> intend the deliberate destruction of the cocoa industry,
> and the acquisition of land for some hidden end?[16]

Though in the 1951 election campaign, the CPP had identified
with the farmer bitterness over the agriculture department pro-
phylactic campaign, in office it soon accepted the judgment of

rts that no other control measure was possible. A measure
compensation was introduced, amounting to 4 shillings per
tree, and 2 shillings per year over three years toward replanting
costs, but this was far from satisfactory to cocoa farmers.

THE INDEPENDENCE YEARS

Despite the difficult moments in the 1950s, Ghana greeted
independence on 6 March 1957, with enormous optimism. The country had $450 million in external assets, cocoa prices remained
relatively good, and there was confidence in the prospects for
moving beyond the monocrop economy. Nkrumah and the CPP appeared
to enjoy solid support, and prospects for political stability
seemed excellent.

The Nkrumah period, from 1957 to 1966, divides into two
quite distinct phases: the consolidation of one-party rule, with
an economic strategy on the whole conforming to the canons of
Western development economics (1957-61); an ambiguously socialist
phase, with growing personalization of power (1961-66).[17] Despite the contrast in ideological emphasis, the attributes of
the regime in its phase two form were all clearly present in the
initial period. The Nkrumah interest in socialism was long-
standing, however numerous the compromises; the seeds of presidential autocracy were already sown within months of independence, as CPP spokesmen began to warn of the need for a "tempo-
rary benevolent dictatorship" in the face of an opposition that
was "violent, waspish, and malignant."[18]

The opposition did, indeed, embrace indiscriminately any
form of discontent in its action. Shortly after independence,
renewed dissidence in the Ewe regions and ethnic restiveness
among the Accra Ga conveyed a message of the fragility of power.
By 1958, the regime adopted preventive detention legislation to
jail opponents, a mechanism of intimidation which soon became
habit-forming. By 1960, the parliamentary opposition had been
all but emasculated, and Ghana had become in practice a one-party
state. A new republican, presidential constitution was approved
in 1960, with simultaneous designation of Nkrumah as President.

Troubles soon crowded in on the economic front. The momentum of the social and economic development program which was the
central plank in the independence platform led to swift expansion
of government expenditures, which rose from 60.5 million pounds
in 1957 to 113.7 million pounds in 1961. Cocoa prices began to
fall in 1959, then broke catastrophically in 1965 to 90 pounds,
compared to 247 pounds per ton in 1957. Ironically, the crash
of cocoa prices partly reflected the success of Ghanaian production, with an all-time record 1965 crop of 550,000 tons induced
by heavy planting in the 1950s encouraged by relatively good
prices, and a slowing of swollen shoot.

By 1960, the government began running growing deficits. The

comfortable sterling reserves of 1957 had virtually disappeared by 1961, to be replaced by the dependency-generating entanglement in international debts. Import licensing, which quickly became a viper's nest of nepotism and corruption, had to be imposed in 1961. Inflation began to take off, reaching 17 percent in 1964, and 40 percent in 1965.[19]

By 1961, Nkrumah concluded that the gathering storm clouds originated in conspiratorial machinations of malignant forces at home and abroad: neocolonial designs of Western powers to subvert African independence, clearly revealed in the Congo crisis; the inherent character of the Western capitalist economy; the incurably bourgeois attitudes of the civil service, which resisted incorporation into the CPP domain; retrograde aspects of Ghanaian society, represented by recalcitrant chiefs and ethnic interest groups, unconscious handmaidens of imperialism. Moderate policies had failed; only a vigorous assault on the enemies of the new Ghana could succeed, informed by a more socialist perspective. The dramatic reorientation of the regime was heralded in the April 1961 Dawn Broadcast.

There were some steps toward a socialist society. Some posts in party and state were placed in the hands of militant socialists. An ideological institute was founded at Winnebah to offer senior officials instruction in the doctrines of the new order. A warm relationship was developed with the Soviet Union, whose ambassador became for a time the most influential member of the diplomatic community.[20] The development plan associated with the West Indian economist Arthur Lewis was scrapped in favor of a more ambitious--some said socialist--program. Export-based growth was downgraded in favor of a commitment to industrial development, the promotion of socialist agriculture through state farms, and comprehensive state management of the economy.

The new socialist-flavored commonwealth swiftly foundered. Signs of trouble appeared almost at once, when workers at the major port of Sekondi-Takoradi launched the first major post-independence strike in September 1961,[21] infuriated by an austerity budget which brought a real wage cut. Plots were uncovered in 1961, and a serious assassination attempt occurred in Kulungugu in August 1962, in which senior CPP officials were implicated.

Cocoa farmers became increasingly resentful of declining real prices, the imposition of a special "development tax" on cocoa in 1959, followed by "forced savings" in 1966, and a further 24 percent price slash in 1965. The price per headload (10 pounds) fell from 80 shillings in 1957 to 40 shillings by 1965, while inflation over this period was 100 percent. With real urban wages in 1965 only half of their 1960 level, one can well understand why the worker and farmer, presumed beneficiaries of the socialist commonwealth, had become resolutely hostile toward it.[22]

Thus it was that the National Liberation Council (NLC) composed of seven army and police officers, which seized power in February 1966, was greeted with overwhelming enthusiasm. The ills of the country were manifest to them: what required systematic dismantling was the CPP socialist commonwealth which had been fastened atop the enduring apparatus of the former colonial state. A new constitutional order then had to be constructed, close to the original design offered by the withdrawing British, but totally reshaped by the Convention Peoples Party.

In truth, little political dismantling was required, as the once-potent CPP wholly evaporated of its own accord. Support of the civil service was quickly won by uprooting political appointees from the regional service and restoring the bureaucracy to the preeminence which, as spinal column of the colonial state, it had long expected. Chiefs who owed their office to the CPP were destooled. A series of commissions of inquiry were designated to expose the patterns of corruption and malfeasance under the CPP, thus banishing its memory to a final purgatory. The former regime had, indeed, provided the investigators with rich lodes of material.[23]

Preparation of a new constitution began quickly, and by 1968 a constituent assembly was at work, selected on a corporatist basis with seats assigned to diverse statutory bodies, voluntary associations, and interest groups. The debates were infused with liberal philosophy, and an abiding fear of the CPP despotic republic. The new constitution was ushered in by what most felt were honest and open elections in 1969.

The second parliamentary republic was born in optimism and good feeling. Two major parties emerged to contend the elections: the Progressive Party (PP), led by Ashanti intellectual and Oxford don Kofi A. Busia, reflecting some continuities in leadership and areas of support to the old anti-Nkrumah opposition; and the National Alliance of Liberals (NAL), headed by quondam CPP Finance Minister of the Arthur Lewis phase, K.A. Gbedemah. The elections were won by the PP with 105 seats, to 29 for the NAL, with 6 in the hands of minority parties. Thus Busia succeeded as Prime Minister in September 1969.[24]

The two-year regime of Busia was a profound disappointment; the reborn hopes of what the political kingdom might deliver were once again dashed. Confidence in the capacity of the classic liberal economy to restore prosperity flagged, as the chronic foreign payments difficulties chained the country to a mendicant role. Meanwhile, tales of corruption in high places began again to circulate, often well-founded. Busia succeeded in offending most important groups: civil servants, unions, traders, cocoa farmers, the army, the judiciary. The final straw was a severe devaluation in December 1971 of Ghana currency, the cedi. As the import component of consumption patterns of higher income groups --civil servants and army officers--is substantial, this amounted to a sharp salary reduction. In January 1972, the army moved once again, and General Ignatius Acheampong became head of state, with fellow officers staffing a National Redemption Council (NRC, redesignated Supreme Military Council in a 1976 reshuffle).

The demise of the Busia regime was greeted with satisfaction, though few could now believe that a dramatic improvement in the conditions of life was likely. For the moment, it appeared that only the soldiers could tackle the tasks of governance, and there was no assurance that they could do it very well. Acheampong at once rescinded the entire Busia devaluation, although not the accompanying measures intended to cushion the devaluation. Since that time, the overvalued cedi has been under constant pressure, with a gradually intensifying inflation. By 1975, hyper-inflation set in: between May 1976 and May 1977, the consumer price index rose 112 points. Shortages of basic commodities have become endemic, and a black market economy has become institutionalized. The porous borders and high profits of smuggling make it impossible to control, particularly since the neighboring states of Togo and Ivory Coast have solid currencies and well-supplied markets. An indeterminate but substantial fraction of the cocoa crop is now habitually smuggled, which at the extraordinary price levels prevailing from 1973 to 1979 represented a grievous loss for the national economy. Legally exported cocoa fell from the historic high of 557,000 tons in 1965 to 320,000 in 1977. Some of this represented a production loss from aging trees and an apparent shift to food crop production, but a significant fraction was illegal exports. It is interesting to note that Ivory Coast cocoa exports rose from 105.2 thousand tons in 1967 to 201.8 thousand tons in 1974. At 1977 prices, in the quite plausible hypothesis that 100,000 tons were smuggled, the foreign exchange loss to the state was $448,000,000. The Acheampong regime had tried to offer more attractive prices to growers, raising the fixed price eight times in five years from 1972 to 1977. But inflation undid the gestures; since 1959, the real value to cocoa producers of the government price has consistently been the lowest in West Africa.

Beyond the smuggling phenomenon, however, colossal losses began to occur through skimming of cocoa proceeds at the top. Ghanaians found it mystifying that the country could find itself in such desperate economic straits at a time of phenomenally high cocoa prices which in 1977 stood at more than 10 times the 1965 level. By 1977, rumors abounded of manipulation of Cocoa Marketing Board transactions, suspicions reinforced by the cessation of published reports by the CMB after 1973. By 1978, West Africa cautiously drew attention to a discrepancy of 500,000,000 pounds between reported Ghana revenues on its officially exported cocoa, and the prevailing world market prices. Much of this colossal loss had been diverted to the private fortunes of Acheampong and some of his collaborators. Acheampong, a man of no known wealth at the time of his seizure of power, managed to squirrel away a fortune estimated by some at $100 million in six years of rule.[25]

Signs of malaise were multiplying from 1974 on, as livelihoods for most deteriorated. Fellow officers ousted Acheampong in July 1978, promising a return to civilian rule by party-based elections. In June 1979, on the eve of the elections, a junior officer junta led by Lieutenant Jerry Rawlings suddenly intervened, with the announced purpose of purging the top army ranks before power was returned to the civilians. After summary court martials, eight top officers, including three former heads of

Table 9.3

COCOA PRODUCTION AND PRODUCER PRICES, 1957–1977

Year	Cocoa Exports (000 metric tons)	Producer Price (shillings/headload)
1957	264	80
1958	207	72
1959	255	72
1960	317	60
1961	430	60
1962	410	54
1963	422	54
1964	436	50
1965	563	50
1966	416	40
1967	381	65
1968	422	70
1969	339	80
1970	414	80
1971	392	100
1972	464	120
1973	420	120
1974	350	150
1975	377	150
1976	397	200
1977	320	

Source: R.A. Kotey, C. Okali, and B.E. Rourke, The Economics of Cocoa Production and Marketing (Legon: University of Ghana, 1974), p. 275; Commodity Yearbook, 1958–1978.

state, were executed--a deep shock in a country where politics were frequently turbulent but almost never lethal. These killings were a fair measure of the depth of the political abyss, as the top of the state apparatus had degenerated into a siphon for illicit enrichment by the ruling military clique.

Evidence accumulated of serious deterioration of the cocoa sector. New plantings fell off sharply from 1960; the trees and the farmers are aging. Various measures, especially the expulsion of many aliens in 1967, have raised labor costs while prices remain discouraging. Farmers have since World War II consistently received less than 60 percent of cocoa's export value, though marketing costs accounted for only 6 percent of f.o.b. returns in the early 1950s, and 14-15 percent in the 1960s. By 1965, the return to farmers on cocoa was only one-third of the 1957 level. In 1973, calculations showed that rice, cocoyams, and yams all had a higher return to the farmer than cocoa, essentially because cocoa was heavily taxed while alternative food crops escaped direct levies.[26] However urgent diversification into food crops may be, Ghana cannot afford the continued decline of its cocoa industry.

Like Uganda, Ghana faced, under the best of circumstances, a grim struggle for recovery in the 1980s. The easy optimism of the dawn of independence has long vanished, replaced by demoralization, cynicism, and even despair. The fairly conducted 1979 elections did produce a clear electoral triumph for Hilla Liman, which gave a faint glimmer of hope. But the atrophy of both polity and economy had gone so far that only a long period of patient convalescence could bring the country back to health. Against this setting, we may turn to an examination of the rise and demise of the Ghanaian cooperative movement.

10.   Cooperatives in Ghana: From Genesis to Initial Dissolution (1928-1961)

The cooperative movement in Ghana was strikingly different from that in Uganda. In the first place, cooperatives played a different role in the conflict over political resources. Whereas the Ugandan cooperatives experienced relative political success, the Ghanaian cooperative movement was identified with unsuccessful political forces. Also, the cooperative policies of successive Ghanaian governments were dissimilar, marked by oscillation between support and hostility.

In this chapter, we examine the cocoa marketing structure and its effect on the cooperative movement; colonial administrative policy toward cooperatives and its impact on the CPP government; cooperative attitudes toward the national political leadership and their consequences for government cocoa marketing policies. Here, we consider these issues for the period from their first creation until their initial dissolution by the Nkrumah regime.

From the early colonial days, Africans were permitted to act as middlemen cocoa traders. It is no coincidence, therefore, that the cooperative movement had its beginnings there. Just as African entrepreneurs responded to the opportunities in cocoa marketing by starting small marketing firms, so too cocoa farmers --themselves entrepreneurs--responded by starting cocoa marketing cooperatives. With no official or de facto restrictions on African participation at the local level, farmers and traders could channel their initiative and drive into making their cocoa trading efforts succeed.[1]

Commodity trading was not a new phenomenon in the Gold Coast. A long tradition made local citizens aware of the dynamics of supply, demand, and price, so it was not surprising that

the African Gold Coast cocoa marketeers brought commercial fi-
nesse, probity, and skill to trading.

Ghanaians had a thorough knowledge of the commodity.[2]
They knew the production process and requirements, the vagaries
of the weather and the soil, the timing, and the limits on production. They also understood their market and the constraints
on moving cocoa from the farm to the buying station. They
shrewdly calculated which footpaths and roads to use to reduce
haulage time. They knew how and where to recruit headcarriers
and how much to pay them. In short, they possessed knowledge of
the local market conditions few Europeans could hope to acquire.

Because cocoa marketing requires little capital, Africans
could become traders with relative ease.[3] Nor did the commodity require special processing equipment or techniques as cotton
and coffee did. After the producer sold them, the cocoa beans
required virtually no additional processing before export.[4] It
had merely to be bulked into loads suitable for head-carrying
from the village to an up-country company buying station.

The European trading firms which dominated cocoa export
trade encouraged development of the African-controlled local marketing systems,[5] concentrating their efforts on the more profitable international commerce. Trading firms like the United
African Company and the United Trading Company provided seed capital to aspiring African traders in order to build a network of
agents loyal to the firm and thereby ensure a reliable source of
supply. For the same reasons, the larger African traders provided initial finance for smaller traders. This network became
a complex maze of petty buyers, transporters, and brokers. By
the 1920s, colonial documents cite with amazement the mercantile
skills displayed in the rural African cocoa trading system.

Thus, the economic environment in which Ghanaian proto-cooperatives first emerged in the 1920s was remarkably different from
that in Uganda where Asians dominated cotton and coffee marketing. In Ghana, cocoa farming itself was an African undertaking,
and a host of socially innovative structures and processes had
developed to resolve the problems of land, labor, and capital required for cocoa production. A dense Ghanaian mercantile rural
infrastructure had grown in symbiosis with the spread of cocoa
farming. It was, thus, in an environment of burgeoning petty
rural capitalism that cooperativism was born in Ghana.

The germ of the first proto-cooperatives may be seen in the
"companies" that formed in various parts of southern Ghana to establish cocoa farms. Ghanaians were discovering the advantages
of pooling efforts to negotiate land rights and to share the burden of the initial six to seven year period when heavy labor was
required for clearing, planting, and tending the young trees before they yielded a salable crop. In the 1920s, farmer buying
associations began to appear, extending the principle of partnership to marketing.

The colonial administration sought to convert these grass-roots associations to its own purposes by absorbing them into an official cooperative structure in 1928. Their major objective, at that point, was to transform the buying associations into agencies for enforcing higher quality control, because the declining standards of Gold Coast cocoa were causing serious problems with overseas customers. Thus conceived, however, cooperatives were unattractive and many farmers simply opted out at that stage.[6] In effect, the friendly embrace of the administration retarded the evolution of the movement. The farmer associations that remained in the official cooperative movement fought to change the colonial administration's cooperative policies and priorities for the next decade and a half. When they finally succeeded, cooperative commercial marketing had advanced little from the days of the informal farmer associations.

DEVELOPMENT OF THE COCOA COOPERATIVES

During the 1930s, the Department of Agriculture treated the cooperatives as an experimental concept and was disinclined to surrender any of its administrative control of the cooperative apparatus to the farmers. The Department made little effort to promote expansion of the cooperative societies. Its success criteria for cooperatives were grading standards and tonnage marketed.

At this time foreign buying firms were in full command of the cocoa market. They set prices, and used their control of capital to enforce their hegemony. Consulting informally among themselves, the large firms fixed the prices they were willing to pay the large African traders, who in turn paid lower prices to their marketing agents. The colonial administration viewed British capital as a means of controlling a rural economy otherwise dominated by Africans. African gestures toward direct exporting were viewed with disfavor.[7]

By the mid-1930s, the Agriculture Department recognized the need to make its flagging quality control program more attractive. They decided to support improvement of cooperative services as an incentive for farmers to join, on the condition that the farmers bring in only high-quality cocoa. The effectiveness of these services ironically weakened Department supervision over time; the enhanced membership loyalty strengthened the cooperatives, and permitted them to become more assertive on their autonomy.

By 1936, the cooperatives began to offer additional services (short-term loans, bonuses, storage) that made them more competitive with the buying agents. The novelty of these new marketing services attracted new members and, at least for the next few years, the cooperatives experienced modest growth. They tried to strengthen their still small portion of the market by amalgamating into unions, an initiative the government did not resist, even though it was apparent that their consolidation weakened the Department's control.

A newfound assertiveness among the farmers and a desire to loosen the Department's control over the cooperatives became evident in 1937-38, when the cooperatives played a leading role in the "cocoa holdup," whereby farmers withheld cocoa from the market for months in a concerted effort to drive prices up to their previous levels. The cocoa holdup was sufficiently dramatic to induce the colonial government to appoint a commission of inquiry. The commission report was hostile to the cooperatives; while it suggested conciliatory moves toward farmers on production and pricing, it did not regard the cooperatives favorably either as a vehicle for representing farmer interests or as a marketing organization.[8]

The outbreak of World War II prevented the abolition of the cooperatives recommended by the Commission. Their report was shelved because the measures adopted to cope with the war effort had completely changed the complexion of cocoa marketing. For the cooperatives, the war measures were a blessing in disguise: the short-term costs were high but in the end the cooperatives used the new marketing policies to break the administration's hold on the movement.

The government imposed stringent economic controls to prevent the collapse of the colonial economies as a result of war dislocation. For cocoa, this took the form of production and price stabilization policies. A cocoa market that had previously been semi-open was now brought under complete regulation--regulation that has persisted to this day.

Government-fixed cocoa prices were to be the longest lasting war measure. The machinery for implementing fixed prices, the Produce Control Board, later renamed the Cocoa Marketing Board (CMB), became a permanent feature of the postwar political economy. The other war measures--the internal two-tier cocoa marketing agents, and internal cocoa quotas (explained later)--lasted only through a few postwar reconstruction years, but their importance to the development of cooperatives cannot be overstated.

The government adopted a fixed producer price policy in order to stabilize production at 1930 levels and to maintain the viability of the colonial economies at 1939 levels. The old floating producer prices were scrapped in favor of a fixed price, but one set well below the 1939 average world market price for cocoa. The difference between the price paid the farmer and the world price (minus the Board's administrative costs, agents' marketing expenses, and a small profit margin for the marketing and exporting agents) went into a reserve fund to be used if wartime world market cocoa prices fell below the fixed price paid the farmer.

In some respects, the fixed pricing policy made the market

more efficient.   Buying agents could anticipate their cash flow,
schedule shipments, and calculate their costs.   By shipping
larger parcels, they could reduce costs in a way impossible under
the old system.   Previously, costs always fluctuated, forcing
agents to ship whatever they had, regardless of quantity, when
an attractive price was offered.   These changes in the market
also benefited the cooperatives, permitting them to assert themselves as a trading force, compete more aggressively for membership, and shed some of their image as an instrument for quality
control.

The government was anxious to limit production to 1939 levels, because higher levels, which the government would be obliged
to buy, might disrupt its wartime budgetary support plans.   Unable to influence output by controlling production, the administration sought to do so indirectly by turning again to the marketing system.

The first instrument for controlling output was the "A" and
"B" shipper designation, a two-tier designation of internal marketing agents.   This measure did two things: it institutionalized
the British private trading firms' dominance of the internal market and it limited the government Marketing Board's contact to
highly capitalized, financially secure "A" trading firms (which
all happened to be British).   An "A" firm could sell to the Marketing Board.   "B" traders (Ghanaians) could sell only to "A"
firms.   Both "A" and "B" firms had quotas but the "B" firms received a lower price for their cocoa than "A" firms did.   For
purchases in excess of the quota, the Marketing Board paid the
agents less than the fixed producer price.

The transparent inequities of the two-tier system provided
the occasion for the cooperatives to become a political voice.
Indignant at the preferential treatment for British firms and
reinforcement of British hegemony, the cooperatives agitated for
equal status, and did not stop until they were recognized as marketing organizations entitled to the same privileges and opportunities as the British firms.

The colonial administration's other policy instrument for
stabilizing cocoa production was the use of buyer quotas for all
firms.   The "A" firms were given larger quotas, which threatened
the financial position of the cooperatives.   The quota policy
therefore became another factor in turning the cooperatives into
a voice of protest, and spurring their drive for recognition as
a marketing agent.

As the war drew to a close, it became apparent from official
remarks and actions that the wartime cocoa marketing measures
would probably continue into the postwar reconstruction period.
Recognizing the likelihood of the Marketing Board's continuing
presence, African cocoa traders and UGCC leaders agitated for
representation on the Board along with the colonial functionaries and European company men.   The agitation began in 1944 and
reached its zenith in 1946.

The increasingly vocal criticism by African agricultural
interests did bring an end to institutionalized discrimination
against Africans in cocoa marketing after World War II. African
representation on the Marketing Board, which before the war would
have been almost unthinkable, came because colonial policy increasingly recognized the legitimacy of the nationalist aspirations. Beginning with the 1947 cocoa season, the Marketing Board
dropped the quota policy and replaced it with a licensing system.
Under a new system, the Cocoa Marketing Board issued licenses to
buying agents whom the Board deemed "worthy buyers" (designated
Licensed Buying Agents, or LBAs). LBAs under the new system
would sell directly to the Board. They would also be entitled
to various benefits, such as CMB cash advances and supplies of
needed marketing materials. Armed with licenses, and freed from
official discrimination, the cooperatives could now compete.
Immediately cooperatives increased their purchases by one-third
and exceeded a 10 percent share of the total market for the first
time in their history.

However, the cooperatives had to contend with one other
mechanism of government control that in the long run compromised
the cooperative movement and contributed to its demise in 1961:
the Department of Cooperatives. This Department was formed in
1944 to relieve the Department of Agriculture from its cooperative responsibilities, which had become burdensome at a time when
the spread of swollen shoot disease was becoming worrisome.

The Cooperative Department was formally charged with supervising, auditing, and registering cooperative societies, and ar-
bitrating disputes between them, in other words, as an agency of
"institutionalized suspicion." These powers had been legally
defined in the 1939 Cooperative Ordinance. The new Department
quickly became a formidable instrument of administrative control.

From the outset, the Department faced a dilemma. Too tight
a control would probably trigger withdrawal of members in the increasingly politicized postwar atmosphere. A marked decline in
cooperative growth, or worse, their collapse, would deprive the
new Department of its raison d'être. Accordingly, the Department
presented itself as an ally of the cooperatives, a posture not
inappropriate to bureaucratic survival instincts.

Its development agenda for the cooperatives was conservative
in every sense of the word. The Department wanted only moderate
expansion of the cocoa marketing cooperatives and limited diversification into new activities. To enforce its policy, the Department adopted a stance of close supervision, coordination, and
control. Although they were ambivalent toward this bureaucratic
ally, the cooperatives postponed whatever hopes they had for
breaking totally away from government control. They focused instead on using the Department to their best advantage.

The Department wanted a moderate increase in the number of
societies and unions, but only after the existing ones were
clearly self-supporting and profitable--and as of 1944 most of

them were not. The Department imposed a ruthless consolidation of societies and unions. First, the Department determined that a society must market at least 100 tons of cocoa per season to achieve reasonable profits. Using the 100-ton figure as a minimum standard of efficiency, the Department forced "inefficient" societies to merge with larger ones, or for that matter with each other.[9]

The Department's consolidation exercise, started in 1944-45, was successful. The number of operating societies was reduced, and society purchases generally began to exceed the 100-ton mark. By showing that a farmer-run, profit-based marketing institution could work, the consolidation program pursued for the next 15 years enhanced the appeal of the cooperatives to nonmember farmers.

As part of its program to expand and integrate the cooperatives, the Department organized two other new projects: the Cooperative Bank and consumer cooperatives. The new venture in cooperative banking was eminently successful; the consumer cooperatives were not. Formed in 1947 and in operation until the government proscribed it along with the rest of the cooperative movement in 1961, the Cooperative Bank successfully provided, among other things, financing for union and society cocoa purchasing requirements, and loans to unions and societies for inventory and capital improvements. The Bank was also largely responsible for spreading financial services throughout rural areas that commercial banks had long avoided. Besides supplying members with a whole range of banking services and financial advice, it accepted deposits from individual farmers. In this way, each union and society acted as a branch of the Cooperative Bank. Over time, the union and society "banks" replaced the time-tested but precarious "stuffed mattress" method of keeping money safe, while introducing an interest-bearing investment and savings opportunity to small farmers.[10] By making cocoa purchasing advances available to societies and unions, the Bank facilitated reasonable short-term loans to individual farmers, an attractive service for cocoa farmers who faced perennial cash-flow problems and an alternative to private moneylenders who charged exorbitant rates.[11]

The consumer cooperatives were organized in 1948, with the double aim of providing the urban population with basic consumer stores, and supplying the cocoa cooperatives with agricultural supplies. Inspired mainly by European precedents, the consumer cooperatives were adapted very little to African economic conditions and within a few years were clearly a failure: managerial expertise was woefully lacking; the urban economy already supported a multilayered network of traders and trading establishments with which the nascent consumer cooperatives were unable to compete.

The success of the cocoa marketing cooperatives and the Cooperative Bank overshadowed the failure of consumer cooperativism. The period from 1948 to 1961 was clearly the "Golden Age" of the cocoa and banking cooperatives. Expansion of the cocoa cooperatives was unmatched by any preceding period, despite the Department of Cooperatives' tight control over the movement. The cooperatives' share of the market grew from 10 percent of total national cocoa purchases to 30 percent in 1960-61, making the cooperatives the largest single buying agent in the market for a time. At the same time a number of farmers recognized that the cooperatives were an attractive investment arena.

One may ask what limited cooperative appeal to a third of the total market and a sixth of the total cocoa farming population. Partially, it was the social composition and orientation of the cooperatives' membership: the larger, more entrepreneurial farmers found the societies attractive. The cooperatives were strongest in the Ashanti and Brong areas where communal village organizations were strong. In addition, the cooperatives came into conflict with the nationalist movement. More precisely, they avoided active involvement in the political struggle for independence, despite appeals and even threats from the nationalist leadership. If one could attribute any political outlook to the cooperatives during the period of nationalist agitation (1948-57), it would be one of indifference occasionally tinged with hostility. In contrast, the CPP was anxious to use cocoa marketing structures as instruments in the political struggle.[12]

Nkrumah was unenthusiastic about the cocoa marketing cooperatives, not only because they were avowedly nonpolitical, but also because they had the same social and political conservatism he saw in the emerging Gold Coast opposition to his political activism. Even had a CPP-cooperative alliance of convenience been possible, Nkrumah seemed convinced that it would have been rife with problems.

Nevertheless, Nkrumah recognized cocoa marketing as the key to the cocoa sector and saw the importance this could have for political mobilization. If the party could organize cocoa marketing, it could build trust with the rural farmers. The cooperatives were politically neutral; the members seemed more socially and presumably politically conservative. The cooperative movement adhered strictly to the Rochdale principle of political neutrality, viewing itself as a service organization for providing its membership with the economic and social services set down in the organization's charter and bylaws. The CPP detected an undercurrent of hostility among smaller cocoa farmers to the entrepreneurs of the cooperative movement, and sought to build on this as a means for penetrating the cocoa sector. To mobilize support for nationalist goals, Nkrumah needed to get a foothold in the cocoa sector. It was in prison in 1949, with fellow inmate Ashie Nikoi, that he planned his first venture in organizing the cocoa trade. Released from prison late that year, Nikoi launched the

Gold Coast Farmers Congress, which for a short time had informal ties to the CPP. However, the Gold Coast Farmers Congress appealed to the same constituency as the cooperatives, the farmer-traders. Nikoi soon cooled on Nkrumah, and his organization did not become a major factor in the cocoa trade.[13]

In 1952 Nkrumah, by then leader of government business, launched the Cocoa Purchasing Company (CPC), a cocoa buying agent that was to become the nemesis of the cooperatives. The brainchild of Nkrumah and Martin Appiah Dangua (later Secretary General of the United Ghana Farmers Council), the CPC was established as a wholly owned subsidiary of the Cocoa Marketing Board, and was led by a mercantile henchman of the CPP, A.Y.K. Djin. CPP spokesmen labeled the CPC its "atomic bomb" because of its potential for strengthening the CPP political machine in the cocoa areas. Supplied with capital by the CMB, the CPC generously though selectively distributed loans to CPP supporters. The CPC purchased 6 percent of the crop in the 1952-53 season, then 18 percent over the next three years. But within a few years, it was bankrupt, riddled with scandal, and a major source of resentment in Ashanti cocoa areas because of the CPP's transparent political manipulation of cocoa patronage.

In 1953, the CPP established another farmer organization, the United Ghana Farmers Council (UGFC, known at first as the Gold Coast Farmers Council and later as the United Ghana Farmers Cooperative Council, or UGFCC). The UGFC was at first a virtual subsidiary of the CPC. However, the UGFC continued to function after the CPC collapsed; indeed, it absorbed much of the CPC staff. Its fidelity to the CPP and to Nkrumah cocoa policies was a heavy burden; after the 1954 elections, it campaigned for a four-year price freeze, and for a sharp rise in export duty. Obviously it was not easy to persuade farmers—large or small— that their best interests were served by a de facto cut in income.[14] However, in taking over the cocoa trading functions of the CPC, the UGFC did attract a following. Its share of the market grew from 6.5 percent in 1957-58 to 23.8 percent in 1959-60.[15] Although the UGFC did not provide the structure or services the cooperatives provided, it seemed to fill a real need in the rural economy: providing African-owned marketing services without requiring share capital or membership participation.

The UGFC was organized hierarchically, with secretary-receivers appointed to run the village societies. Democratic participation and control were not salient features of the UGFC, but the village societies did have three-man village advisory committees, composed of the chief farmer (the CPP title for the largest farmer in the village) and two other appointed representatives.

The UGFC's market position between 1955 and 1961 improved, largely because of the CPP's national political leadership. The CPP increased the Council's visibility and power by giving it a

seat on the Cocoa Marketing Board and by including it in the top
councils of government. In 1957, the UGFC was declared sole political representative of the farmers, provided an annual 100,000
pound subsidy by the CMB, and equipped with a 100,000 pound
building.[16] These gestures certainly made an impression on
many farmers.

To establish UGFC (and thus CPP) control over cocoa marketing, the UGFC had to displace three adversaries: the British cocoa trading firms (especially United Africa Company and Cadbury),
private African cocoa brokers, and the cooperatives. Ironically,
the multinational giants proved to be the weakest foe. In 1959,
the UGFC launched a bitter rhetorical offensive against foreign
cocoa traders. The same year, the UAC redeployed its interests,
announcing its withdrawal from cocoa trading in favor of various
other activities (sawmills, a brewery, a soap factory, and Kings-
way stores).[17] Cadbury, whose interests were more concentrated in cocoa products, was more resistant, but by 1961 concluded that takeover was inevitable, and chose to cooperate with
it. From 1961 to 1963, Cadbury supplied professional staff to
the UGFC.[18]

The independent African brokers had been a declining factor
since the war: relatively few had been recognized as Licensed
Buying Agents under the postwar legislation, and in the 1950s
they handled only 2 percent of the crop. However, with the foreign firms gone, the African brokers were doubtless a threat, and
the UGFC launched a vociferous campaign against their purported
exploitation and cheating (including the comingling of good and
bad cocoa), and called for a noncapitalist method of cocoa marketing. The private brokers had little political support; indeed, many were Nigerians (mainly Yoruba). They offered little
effective resistance to the UGFC drive for monopoly.

The cooperatives provided more formidable opposition. With
a number of friends among CPP backbenchers as well as United
Party members, they had significant backing in Parliament. The
Department of Cooperatives remained a bureaucratic ally, even
though its influence with the new CPP government was limited.

In 1959, the government announced a plan for unifying the
cooperative movement, which effectively meant bringing the cooperatives under the UGFC's wing. The cooperatives rebuffed this
initiative, pointing out that the UGFC was not constituted in
accordance with cooperative principles. The following year, the
Ashanti and Brong Ahafo societies broke away to form their own
body, ABASCO (Ashanti and Brong Ahafo Cooperative Organization).
This split in the cooperative movement--ostensibly triggered by
southern domination of the cooperative leadership--provided a
timely opening for the UGFC. In a major breakthrough, it negotiated the incorporation of ABASCO into its ranks later in 1960.

This left only the rump of the cooperative movement, which
was subdued by fiat in April 1961. The government announced that

the UGFC was to be awarded a trading monopoly in cocoa, as part
of the general shift toward a noncapitalist pathway announced by
the Dawn Broadcast. At the same time, the Department of Cooperatives, which remained an irritating bureaucratic spokesman for
the classical Rochdale principles of cooperation, was abolished.[19] Not only were the cooperatives forced out of the cocoa trade, but their quite considerable assets were simply confiscated. Farmers lost $3 million in share capital and savings
deposits, as well as substantial capital gains realized in the
form of appreciated fixed assets and interest-generating reserve
funds. The 50,000-odd cocoa farmer cooperators were powerless
before the CPP juggernaut. However, their bitterness contributed
to the disaffection that ultimately sapped the once-solid popular
foundations of the Nkrumah regime.[20]

CONCLUSION

     The CPP government liquidated the Ghanaian cooperatives be-
cause the party viewed them as a vestige of colonial rule, a
striking contrast to the strong support cooperatives received
from the Obote government in Uganda. The accusation was not totally unfounded. The colonial administration had tried to supervise and control the movement from the time of its formal incep-
tion in 1928. However, the farmers had eventually wrested control of the cooperatives from the colonial administration, and
then used the administration to serve their own ends.
     The colonial administration had lost this struggle for domi-
nance largely because its goal of using the cooperatives as an
instrument of cocoa quality control could be realized only if the
cooperatives were viable organizations. In achieving commercial
viability, the farmer-run cooperatives redefined their needs and
goals. Instead of being an instrument for quality control, they
evolved into a vehicle through which the relatively successful
commercial farmers who dominated the membership tried to influence the cocoa marketplace. The cooperatives offered competitive prices, more marketing services, and an arena for farmer
investments; this quite efficient performance made them a major
force in cocoa marketing.
     However, in winning the battle with the colonial administra-
tion and then developing a close working relationship on their
own terms with the Department of Cooperatives in the final years
of the colonial administration, the cooperatives unwittingly acquired a more dangerous adversary in the CPP. The collision
course between the cooperatives and the CPP was perhaps inevitable. The CPP sought full hegemony over society and economy, and
profoundly distrusted all opposition, real or potential. After
a thirty-year struggle for autonomy, the leaders of the cooperatives were not receptive to a few half-hearted CPP gestures at

coopting them; they were thus automatically viewed as at least latent opponents. Further, the CPP found autonomous centers of socioeconomic power particularly distaseful in this critical sector of the economy--although even at their peak the cooperatives never dominated the cocoa market.

Finally, the CPP ultimately depended on the taxes extracted from the cocoa sector to finance its development plans, and to enhance the state apparatus it believed necessary to transform society. While high world prices and swelling output during most of the 1950s made this strategy workable, a sharp decline in world prices in the late 1950s and early 1960s meant that heavy levies on the cocoa sector were keenly felt and bitterly resented by cocoa farmers, and not only by those who belonged to cooperatives. The CPP's confrontation with the cooperatives from 1959 to 1961 was the opening act of a broader crisis that finally destroyed the regime's relations with the cocoa sector.

The forced liquidation of the cooperatives in 1961 destroyed what in many respects had become a quite useful instrument for facilitating cocoa marketing. The cooperatives were not, as we have seen, a really egalitarian or redistributive vehicle. Their leadership, while it did not mirror traditional hierarchies, did reflect the new rural stratification born of differential commercial success in cocoa cultivation. Even here, it is important to keep the degree of inequality in perspective; in 1963-64, while about half the cocoa sales did come from about 10 percent of the farmers, only about 5 percent of the approximately quarter million cocoa growers sold over 200 loads, thereby grossing over 500 pounds. The really prosperous cocoa capitalist farmers numbered no more than 5,000, whose farms averaged only 50.2 acres, and part of whose income came from trading.[21] Their really major gains came in the 1950s; by the 1960s, they no longer stood out as unusually prosperous Ghanaians.

If not egalitarian, the cooperatives at this epoch were certainly quite efficient. Operating with a high degree of autonomy, they had carved out a substantial market share for themselves in active competition with the European cocoa houses and private African traders. They were certainly no more inegalitarian in operation than the alternative marketing vehicles which they fought, or the CPP-imposed UGFC. Their success in winning a large share of the market by their own efforts stands in sharp contrast to the Uganda situation, where the cooperative movement was closely sheltered by the state, and achieved its leading position in cotton and coffee primarily by government intervention in its behalf.

11. Reorganization and Breakdown: The Ghanaian Cooperatives from 1966 to 1977

RESURRECTION OF THE COOPERATIVES

The demise of the Nkrumah regime in February 1966 provided the opportunity for a resurrection of the cooperative movement. However, the cooperative experience during the next eleven years proved disheartening for supporters of the cooperative concept. After a brief period of euphoria and enthusiasm, cooperatives fell into a period of decline. Saved from total collapse by a change of government in 1972, the cooperatives briefly revived amidst renewed enthusiasm, only to experience a new sequence of decay, virtual collapse, and final abolition in the spring of 1977.

Within a few days of the overthrow of Nkrumah, the leaders of the coup, General Ankrah and Colonel Kotoka,[1] announced the abolition of all Convention Peoples Party organizations, including the United Ghana Farmers Cooperative Council (UGFCC, formerly the UGFC).[2] Cooperative records show that the day after the government seized power, the former cooperative leaders, having been banished from public life after the takeover of the cooperatives in 1961, got in touch with one another and called for a meeting to discuss reestablishing the cooperatives.[3] For the next six months, the former cooperative leaders met many times to plan a revival strategy. With their former regional, district, and grass-roots leaders mobilized, the cooperatives represented a formidable lobbying presence, as well as a marketing organization ready to get back into business at a moment's notice.

The collapse of the CPP and the total discrediting of the UGFCC provided a new opening. Even before the coup, the UGFCC was on the defensive. Its spokesmen, in rebuttal to increasingly bitter attacks, pointed to its success in moving the cocoa crops, and--ironically--in achieving the colonial administration's goal of improving the quality of the cocoa. The percentage of Grade 1 cocoa exported had increased from 73 to 99.8 under the UGFCC monopoly (quality rating depends on the number of defective beans found in a given consignment). This was accomplished by making buying standards severe. The value of this achievement was lessened by the failure of the international market to pay much of a premium for the higher quality; as a result, the farmers took a net loss, as they still received the same price, but required more labor to pick over their beans.[4]

Further, despite the UGFCC's claim to efficiency, its dealings were increasingly vexatious to farmers. The secretary-receivers were accused of corruption, private profiteering in transport and in cocoa shed construction, delays in crop buying and cash payments, overhiring, and chiseling on weights. There can be no doubt that the private profits that had formerly accrued to the brokers and foreign firms now benefited the UGFCC bureaucracy, not the farmers.

Equally irritating was the transparent contradiction between the UGFCC's theoretical role as voice of the farmers, and its actual function as an instrument through which the regime could control the cocoa sector. From the outset, "democratic centralism" had prevailed. At the first UGFCC delegate's conference in 1955, it was decided that the leaders would all be elected for life, using the dubious analogy of the enstooled chief. In local societies, elections were rare, and officers and bureaucrats were wholly insulated from local pressures, subject only to the central hierarchy in Accra. The UGFCC dutifully claimed fervent farmer support for the various sacrifices imposed on cocoa farmers (a development tax, forced savings, and price cuts). In 1965, the now docile CPP National Assembly duly noted the patriotic support the UGFCC had articulated for the final price slash, thanking farmers for "their shining example of sacrifice," which originated in their understanding "that they belong to a Socialist Society in which the interest of the State is paramount."[5]

Beckman, in his superb study of the UGFCC, persuasively identifies the central contradiction between the CPP and the UGFCC as the different social interests they serve: the UGFCC, like its CPP sponsor, was not an organization of farmers, but a quasi-political bureaucracy. Beckman argued:

It was only for the members of the urban-based petty-bourgeoisie with its degree of education that a "non-

capitalist" solution could hold up a credible alternative
as a means of private advance. Only they could compete
for the offices of the new party and state bureaucracies.
The real losers were not the party businessmen who had
fought in the National Assembly for free enterprise and
free competition. Few of these had any private interests
in the cocoa trade. The losers were the farmers-traders
who had handled the bulk of the trade either as brokers
or as cooperators. Recruited mostly from the upper stratum of the farmer community, these were the people with
the most experience in organizing the farmers and those
most likely to turn farmer organizations against the
central government and its heavy appropriation of cocoa
income.[6]

Thus, the National Liberation Council (NLC) was faced with
the immediate need for a new structure for cocoa marketing. Uncertain if it wanted to retain a single buying agency system or
develop a competitive system, the NLC debated the question for
the next five months.[7] There were numerous supporters of a
nonpolitical, single buying agency system, especially among the
top managers of the Cocoa Marketing Board. To propose a solution
to the government, a Committee of Inquiry, known as the DeGraft-
Johnson Committee, met from May through June 1966. Their report
recommended the establishment of a competitive cocoa buying system that would include the cooperative movement, a cocoa marketing board operating agent, and private licensed buying agents.[8]
  Although the report was submitted to government in July, the
NLC delayed announcing a decision until October, only a week before the cocoa buying season officially opened. From July to
October, supporters of the single buying agency system lobbied
hard to sway the NLC from the DeGraft-Johnson Committee recommendations. They failed.[9]
  A week before the cocoa buying season started, the NLC announced its willingness to allow the cooperatives to reenter the
cocoa market.[10] With their leaders in place and ready to
move, the cooperatives were able to set up operations within
seven days. The leadership literally seized its former offices
and sheds from the Cocoa Marketing Board secretaries who were
acting as overseers of the property until a decision was made on
the structure of the marketing system. The cooperatives lacked
cash to purchase the crop, meet operating expenses, and secure
new equipment, but, under pressure from the NLC, the CMB came
through with the necessary advances for financing the first several weeks of cocoa purchases.[11] With cash available to begin
operations, the cooperatives reentered the cocoa market in October of 1966 with 34 unions, 444 societies, 59,760 members, a
small fraction of their former assets, and a great deal of
optimism.[12]

WHY BREAKDOWN?

Government Policy

After this promising resumption of activity, why did the Ghanaian cooperative movement then break down?

In our judgment, the decline in the effectiveness and efficiency of the cooperatives resulted from an interplay of inadequate finance, poor management decisions, conflicting demands from members, and lack of government support. There was probably some threshold point beyond which atrophy, decay, and cynicism had become too deeply embedded for any action short of starting over to be effective. Once that point was passed, the various factors that had led to breakdown reinforced one another, creating a vicious circle of inefficiency, ineffectiveness, decay, and collapse.

We believe that the cooperatives passed that critical moment some time in 1968. After that it was impossible for the cooperatives to establish an efficient and viable organization without total reorganization. Overexpansion, debt, mismanagement, corruption, misapplication and misappropriation of funds, factionalism, insolvency, and finally an inability to market their produce plagued the cooperatives from 1968 until 1972.

The NLC's policy toward the cooperatives and, generally, toward the cocoa marketing system, can best be characterized as one of benign neglect. The NLC limited its policies to establishing a framework for a competitive system and passing the enabling legislation that would allow the cooperatives to return to the marketing system with the support of a government-sponsored Department of Cooperatives. NLC Decree Number 252 reestablished the Department of Cooperatives with full authority to organize, supervise, audit, and mediate conflicts for the cooperative movement. Though this gave the cooperatives an aura of official sanction, the government played virtually no further direct role in supporting cooperation.[13]

The cooperatives wanted additional government backing. Their demands were simple: to operate; to purchase new equipment; to obtain government-guaranteed medium- and long-term financing; and to get back the assets seized earlier, either in cash or in kind, at current market value. At this early stage, the cooperatives did not demand an official role in, or representation on, government bodies.[14] Nor did they demand government-sponsored small-farmer credit to cocoa producers, or any other system of government-financed short-term loans or advances. The cooperatives publicly and officially welcomed the reintroduction of a competitive system, arguing that it was the best avenue to marketing efficiency, and to meeting producer needs by offering alternative sales outlets.[15]

The NLC resisted the cooperative demands. It refused to

call a government commission of inquiry into the seizure of cooperative assets for two years. The Moore Commission, which was
finally set up in 1968 to study the issue of confiscated cooperative assets, did recommend favorably for the cooperatives.[16]
However, the government did not respond with an official White
Paper outlining its actions on the Moore Commission recommendations until 1970.[17] By that time, the cooperatives were well
on their way to financial and operational decay.

The NLC, and the Progress Party government after it, refused
to provide a government guarantee for cooperative loans secured
from the private money market. The cooperatives had enjoyed such
a guarantee in the pre-1961 period and had never fallen in arrears on principal or interest payments. Nonetheless, the NLC
and the Busia government argued that the cooperatives' financial
position was too weak to warrant the government taking a risk on
their future.[18] These fears were not groundless; after the
first two seasons of operations, the cooperatives were deeply in
debt.[19] But, had the government stepped in early in the first
season of operations, the crippling indebtedness probably could
have been averted. With government backing, the cooperatives
could have secured long-term financing that would have enabled
them to purchase needed equipment and to meet short-term operating costs until the movement was once more financially viable.

The NLC also refused to reestablish the Cooperative Bank.
Close observers of this affair suggest that the Ghana Commercial
Bank had successfully lobbied against Cooperative Bank reestablishment for fear that a new bank would take the lucrative cocoa
financing business away from the government-sponsored financial
institution.[20] The Progress Party government provided the enabling legislation to reestablish a cooperative bank but refused
to come forward with either a long-term loan (once promised to
be 10 million cedis) or equity participation in the institution.[21] Although the bank existed on paper, in reality it was
a nonfunctioning skeletal organization with no loan portfolio
whatsoever until after 1972.

The motives behind NLC policy were not entirely clear;
neither was the policy itself free of ambiguity. NLC first reestablished a competitive marketing system on the recommendation
of the DeGraft-Johnson Committee. A competitive market fit into
the NLC's development scheme: economic competition, with private
enterprise taking on a new role. However, with the Cocoa Marketing Board generating a hue and cry against the competitive system
and calling for a state-controlled single-agent cocoa buying system, the NLC found itself pressured to change policy. As problems grew larger and as the clamor to abolish the competitive
system became louder, the NLC seems to have adopted a wait-and-
see attitude, hoping that a single buying system would emerge on
its own from the internal collapse of the competitive system.[22]
Another reason cited by former cocoa marketing and coopera-

tive officials for NLC ambivalence was the cooperatives' tainted reputation.[23] The defeat endured by the cooperatives in their unequal confrontation with the CPP left in its wake a widespread belief that the movement was a loser, unable to withstand political pressure. This image of sickliness greatly reduced the appeal of cooperatives to the NLC and Busia. Many contend that if the cooperatives had reestablished themselves as a viable and efficient marketing organization, the NLC and subsequent governments would probably have embraced the movement. But many former members and staff were unwilling once more to take a risk on the cooperatives. Without a clear, strong statement of support from the government, and with only a portion of its former leadership, the cooperatives acquired a reputation for being mediocre.

The Progress Party government was more overtly hostile to the cooperatives than the NLC. By the time Busia took power in 1969, the country was well aware of cooperative financial and marketing problems. Although the Progress Party government came into power with a party program supporting reestablishment of the Cooperative Bank, elimination of income taxes on cooperative profits, and a return of cooperative assets, the PP government quickly moved away from the weakening cooperative organization.[24] The government fulfilled only its promise to eliminate income taxes on cooperative surpluses.

For the next two years in office, several influential leaders of government who were known to be close to the Prime Minister expressed public dismay with cooperative management.[25] Having ignored cooperative pleas for financial assistance, the PP government delivered a virtual coup de grace by not reappointing a single cooperative representative to the Cocoa Marketing Board.[26] The Board was hostile to the cooperatives on the grounds that they were mismanaged and financially derelict. Beginning in 1968, the CMB demanded repayment of monies due from the cooperatives against cocoa purchasing advances. When the cooperatives were unable to meet this demand, the CMB refused further crop finance, with the tacit agreement of the government. The cooperatives muddled through on 90-day notes from the Ghana Commercial Bank until a new government seized power in 1972.[27]

The Progress Party government, and Busia in particular, were believed to support the private licensed buying agents over the cooperatives.[28] The PP government wanted the cooperatives to collapse, the argument ran, so that the void might be filled by well-known, private cocoa marketeers who were also major supporters of the Progress Party. The CMB was ordered to relax the registration licensing rules for the private buyers. Although it was well known in cocoa marketing circles that many private licensed buying agents had not put up the capital required to obtain a license from the Cocoa Marketing Board, they were issued licenses anyway--in response, many believed, to pressure from the Prime Minister's office.[29] This mixture of benign neglect and

outright hostility left the cooperatives on the verge of collapse
by the time that the Busia government itself was eliminated in
1972. At that point, the National Redemption Council government
radically reorganized the cooperatives and the cocoa marketing
system, and for a brief moment seemed to bring them back to
health. However, the cycle of decay set in anew, as the state
itself degenerated into an engine of corruption.

Financial Constraints

     The cooperative breakdown reflected in part an incompatibil-
ity of the management system with the resource base. From the
start, the reborn cooperatives lacked the capital necessary to
meet financial and operating expenses. That they never overcame
this initial constraint was partly the result of poor management
decisions at all levels.
     From the first weeks of operations in 1966, the management
at the union and society level adopted a highly questionable
strategy for dealing with this lack of capital: borrowing short
and lending long--that is, diverting short-term purchasing advances to finance long-term expenditures.[30] Capital, in the
form of equity, share capital, member deposits, or reserves or
long-term debt, was simply unavailable during the first year of
operations.
     The pre-1961 cooperatives had an impressive capital posi-
tion, with 12.7 million cedis in assets, including 1.8 million
in cash, 1.3 million in investments, and 1.5 million in properties. The cooperative liabilities had included 1.96 million
cedis in share capital and 1.5 million in member deposits, plus
a reserve fund valued at 1.2 million.[31] This substantial capital position had given the cooperatives considerable discretion
in investments, expenditures, and lending. The old movement had
financed part of its purchases with its own funds. It had had
ready access to outside commercial loans, in addition to the capital made available to it by the Cooperative Bank.
     The post-1966 cooperative movement began operations with
only a fraction of this capital. During the first weeks of operation, cash was virtually nonexistent. Members were unwilling
to contribute share capital until the assets seized by the Nkrumah government were returned. The commercial banks were unwilling to lend the cooperatives long-term financing if the cooperatives would not put up substantial collateral. The cooperatives
had some assets; many of their members had physically seized some
of the properties the Cocoa Marketing Board had been charged with
overseeing until the government made a decision about the return
of assets. At the end of the first year, the cooperative societies had 1.7 million cedis in assets, including 537,567 cedis
in cash, but most of this represented cash advances by the Cocoa
Marketing Board. However, these assets were pathetic by compari-

son with the pre-1961 period. Nor had the cooperative societies
attracted much in the way of liabilities: only 8,000 cedis in
share capital, 3,200 in member deposits, and reserves of 4,500.
The union balance sheet was not much better. Cash assets totaled
389,000 cedis, most of it from Cocoa Marketing Board advances.
Properties were valued at only 75,000 cedis. The only significant amount of capital was 577,000 cedis in member deposits, almost all of them from unions in the Ashanti and Brong Ahafo regions. The Ghana Cooperative Marketing Association (GCMA) had
virtually no investments, properties, share capital, deposits,
or reserve funds, although its balance sheet shows cash holdings
of over 5 million cedis, all of it from CMB advances.[32]
     The cooperative capital position improved somewhat in the
following years, but the operating ratios suggest that most of
the cooperatives were theoretically bankrupt, with current liabilities far exceeding current assets. The unions and societies
survived only by rescheduling their debts, often misapplying purchasing funds or getting additional local financing. Another
practice the societies and unions commonly adopted was to refuse
repayment of cash advances from the GCMA, which in turn had to
repay the advances of the Cocoa Marketing Board. Balance sheet
data are unavailable beyond the 1967-68 cocoa season. From what
data were obtainable, it is our impression that the cooperatives
were totally bankrupt by 1969; by 1972 they owed 6 million cedis
to the Cocoa Marketing Board and 3 million cedis to farmers.[33]
     In 1969, the Moore Commission recommended that the govern-
ment assure the return to the cooperatives of all disputed assets, pay the Ghana Cooperative Marketing Association 1.3 million
cedis for the balance of its assets, and provide a soft loan of
1.5 million cedis. After some delay, the government finally provided a medium-term loan of 1.5 million cedis in 1970. This
helped the movement partially repay the CMB, and eventually enabled the cooperatives to obtain a loan from the Standard Bank.
However, even this government credit was extended on the belief
that the cooperatives had matching capital, which they did not.
     Cooperative management made several poor decisions on the
use of its funds. Significantly, the GCMA management committee
used the first 500,000 cedi installment from the Standard Bank
to finance member bonuses for the 1969 season rather than to purchase equipment and property.[34] Several influential members
of the cooperative movement argued against the decision. When
the management committee refused to listen to their advice, committee members in the Ashanti and Brong Ahafo unions tried to
break away from the GCMA, citing poor management.[35] The management committee assumed that the bonus would generate new membership interest in the movement and lead to significant contributions of share capital; however, these failed to materialize
and the GCMA could show only an additional 500,000 cedi debt.[36]
     At the unions, management committees also often made poor

financial decisions.   One common union practice was to use CMB
purchasing advances to buy trucks.   The committees assumed--
incorrectly, as it turned out--that they could repay the investment within the year because of the savings realized by doing
their own trucking, plus income from outside trucking business.
Similarly, union management committees purchased new scales,
sheds, and tarpaulins in the belief that within the year they
would have generated sufficient surplus to pay back the Cocoa
Marketing Board.

As a result of undercapitalization and financial mismanagement, the cooperatives thus fell into debt.   As debt increased,
the societies and unions defaulted on payments to farmers for
produce.   In so doing, they destroyed any possibility of generating income from farmer contributions.

Overexpansion of the Cooperative Movement

Rapid expansion of the number of cooperative unions and societies greatly contributed to the cooperatives' financial weakness.   New unions and societies broke away from existing ones as
village leaders, aspiring politicians, and frustrated staff members clamored for new units.   Many saw the expansion of the cooperative movement as healthy growth, despite warnings from government staff and international experts brought in to evaluate the
movement.[37]
The overexpansion of the cooperatives hurt their financial
position for two reasons.   First, as each new union or society
was formed, management committees diverted purchasing funds to
equip the new society or union and meet operating expenses.   This
accelerated the cooperatives' indebtedness.   Second, the increased number of unions and societies was not matched by an increased share of the market.   Generally, the new unions and societies were not economically viable and often competed with their
neighboring or parent unions and societies for produce.

The reborn Cooperative Department still estimated that a cooperative society needed to purchase 100 tons of cocoa per season
to be economically viable.   On the average, the cooperative societies reached this goal during the first two years of operation,
even though the number of societies and unions increased dramatically during the second year.   Cooperatives had increased their
share of the market from 28 percent in the first year to 43 percent in the second.   However, with the proliferation of new societies, by the end of the third year the average society fell well
below the 100-ton mark.

The GCMA management committee refused to consolidate the societies.   Despite advice from all sectors, they argued that expansion represented local initiatives with which the committee
did not want to tamper.[38]   By 1970-71, the cooperatives had
expanded into 72 unions with some 1,700 separate societies (reg-

istered and unregistered). At the same time, the cooperatives reached their low point in total market share at 20 percent. Moreover, the per society tonnage dropped to new lows, averaging as little as 21.5 tons in the Volta region, and 33.6 tons for the Central region.

Mismanagement

Mismanagement, such as mishandling of finances, was a major weakness in the post-1966 cooperatives. There were several reasons for this. The GCMA could not gain control over the operations of the unions and societies because of their insistence on local autonomy. Central management was virtually powerless to organize, direct, or control the affairs of the unions and societies. Union and society staffs were poorly supervised by committee members. As the cooperatives fell into debt and failed to pay bonuses because there were no surpluses, members lost interest in cooperative affairs. Management committees felt little incentive to supervise union and society staff members and an authority vacuum developed. With virtually no accountability to management committees or membership, many society and union secretaries fell into a pattern of embezzlement that affected the entire movement.

The organization of the cooperative movement made it easy for a vast system of corruption and misappropriation to evolve. Each tier in the movement--the Association, the union, and the society--was responsible for its own affairs, including costs. Since the unions and the Association were created by the societies, theoretically all power lay at the society level. The Association and the unions could suggest appropriate measures for the societies to adopt but could not enforce their recommendations. Thus, authority flowed in one direction, but cash flowed in another. The Cocoa Marketing Board made advances to the Association, which advanced the cash to unions and societies. With the old cooperative organization, this system worked well, since the society and union committees diligently supervised their own operations and saw to the repayment of cocoa purchasing advances. The post-1966 unions and societies insisted on using cocoa purchasing advances to buy equipment and meet other operating costs, and the Association was the only body legally responsible for the money. The Association did make fruitless efforts to exercise control over local use of funds; many memos circulated, urging the unions and societies to adhere to the rules governing the use of cocoa purchasing advances, but they were summarily ignored. The Association had statutory authority in state legislation to exercise total management control over the unions and societies. However, it refused to adopt a hard line with its member units.

Financial mismanagement put the cooperatives into 8 million cedis of debt. As cooperative debt increased, the Cocoa Market-

ing Board refused to advance more than 20 percent of the weekly
cash requirement for purchasing cocoa. As cash became short, the
societies resorted to giving the farmers chits (IOUs). However,
these represented promises of future payment of money the societies and unions had already spent. The cooperatives had virtually no hope of recouping their losses except to obtain long-term
financing to repay the members for the monies already spent.[39]
     By 1971, the cooperatives were virtually without management
control. Stories abound of society presidents and treasurers unwilling to go to the coop shed for fear of being beaten by farmers incensed about the society's nonpayment for produce delivered. Union and society staff refused to be seen at the society
shed except to receive the weekly cash advances which they would
quickly disperse to those who had their IOUs ready. The staff
would then leave the premises and return to safety.

Factionalism and Regional Conflict

     Another factor contributing to the breakdown of the cooper-
ative movement was the factionalism and regional conflict that
plagued the movement from 1969 to 1971. Considerable energy and
resources went into managing this conflict, which was fought in
public. These unseemly squabbles did little to reassure members
about the future of the cooperative organization.
     An Ashanti and Brong Ahafo secessionist movement, similar
to that exploited by the UGFC in 1960, grew out of their dissatisfaction with the GCMA's management. The Ashanti and Brong
viewpoint was that cooperative unions should make cash advances
only to unions that were not in debt. The Ashanti and Brong
unions (again united as "ABASCO") disagreed with the GCMA's decision that all the unions had to help reduce the Cocoa Marketing
Board's advances. The ABASCO unions had operated relatively efficiently, and were generating considerable contributions of
share capital from their members which they used to meet the cost
of capitalizing their operations and in some cases their operating expenses as well.
     The ABASCO unions were further incensed by a GCMA proposal
in 1970-71 to solve the indebtedness problem by centralizing
management of cooperative finance. The ABASCO unions felt that
centralizing the management of finances avoided the issue of
poor management control. Centralization would punish efficient
unions, reducing their power and authority for the sake of inefficient unions. Moreover, centralized management violated the
spirit of local autonomy, cherished by the ABASCO unions. They
argued that the offending unions, which were also generally located in southern Ghana, should be left to their own fate.[40]
The GCMA management committee wanted to treat all the unions
equally, a position they held to be in the spirit of cooperativism. The Association committee therefore resisted all efforts

by the ABASCO unions either to decentralize control of the Association to the regions or to split the movement into two autonomous factions.

The ABASCO effort to regionalize the cooperative movement
was eventually reduced to an ethnic battle. Accusations of im-
propriety and abuse of office, which filled the columns of the
local papers, were directed against the GCMA President, a south-
erner, and his predominantly southern committee and staff. Finally, much of the southern leadership of the cooperative movement either resigned or were forced from office in a series of
turbulent meetings, to be replaced primarily by Ashanti and Brong
Ahafo candidates. With their own men in power, the ABASCO unions
dropped their demands for regionalization. But much damage had
already been done, and skepticism about the future of cooperatives reinforced.

The new cooperative leadership had only two months to act;
its term of office was cut short by the military takeover of the
government on 13 January 1972. Within eight days of the coup,
the new National Redemption Council government dissolved the GCMA
management committee and installed a military manager to run cooperative affairs. The manager immediately centralized all operations in the hands of a new GCMA management committee, composed
of his staff and himself. The union and society management committees were reduced to advisory boards with virtually no authority over staff. Union and society staff members were directly
accountable to the GCMA manager.[41]

Coping with Mismanagement

Throughout the period of deepening crisis in the cooperatives, nowhere do we find evidence of grass-roots appreciation
for the causes and effects of the cooperatives' mismanagement.
In fact, societies and unions throughout the country resisted in-
stituting measures that would have increased the control of union
and society management committees over staff. For example, union
and society committees refused to require that each society or
union secretary put up a surety bond during their term of employment. The GCMA management committee strongly recommended bonding, which was common practice in government. (Bonding insured
recovery of losses when employees were caught embezzling funds
or taking government property.) The committees also resisted
standardized tests or criteria for employment.

Poor pay was a major reason for indifferent management performance in the societies and, to a lesser extent, the unions.
The average salary for a beginning secretary-receiver was 30 cedis per month in 1973-74, barely above the government's minimum
basic wage for nonagricultural workers. There was little opportunity for promotion, so staff morale at the society level was
low. Our interviews with secretary-receivers revealed little

commitment to the cooperative movement; secretaries saw their positions as a job, nothing more. Low pay and little chance of advancement tempted them to misappropriate cash either by stealing
it outright or by granting themselves advances for personal use.
(Many society staff members believed that they would repay their
cash eventually, or that those who were guilty of taking money
had intended to do so but were unable to.)

At the GCMA, the management committee resisted the idea of
establishing a strong, independent audit unit to verify the accounts and monitor the activities of societies and unions. From
1966 until 1972, Association management was divided over the
question. Although the GCMA's audit section strongly lobbied for
an independent, fully staffed audit unit, other GCMA leaders resisted the idea, apparently fearing that an audit unit would
weaken their own authority. When an audit unit was finally set
up in 1972 following the NRC reforms, its effectiveness was limited by inadequate staff and funding.[42] In the first year of
operations, the audit unit for southern Ghana was staffed with
33 people whose job it was to audit 367 societies and 12 unions.
By 1974, the audit unit had a staff of 70 to audit 1,600 societies and 72 unions. With only limited transportation available,
their job proved impossible. Auditors could not possibly visit
each society monthly, as had been the practice before 1961; in
fact, some societies were audited only once a year. Thus it was
impossible for the unions and Association to monitor society financial activities closely.

BREAKDOWN, REORGANIZATION, AND ABOLISHMENT

From 1970 to 1972, the cooperatives were no longer effective
marketing agents. In scattered parts of the central, eastern,
and western regions, cooperative societies simply ceased to function. In other areas of the country, unions and societies could
not quickly transport the cocoa from sidings because they lacked
cash to pay for transportation. In many cases, they also lacked
cash to pay the farmers for their cocoa. Marketing problems had
reached crisis proportions by 1972 when the NRC took over the
government and instituted a radical reorganization of the cooperative movement.

At the time of the NRC takeover, the cooperatives' estimated
debt stood at 8 million cedis. They owed some 6 million cedis
to the CMB and at least another 2 million to farmers.[43] Under
the tutelage of the new military leadership, the cooperatives began to prosper again between 1972 and 1975. The military government was swift to arrest suspected embezzlers within the cooperative movement. The secretaries were bonded and made directly
accountable to the Association. Corruption and misappropriation
virtually disappeared. The NRC directed the Cocoa Marketing

Board to write off the monies owed by the cooperative movement, and to advance sufficient cocoa purchasing funds to carry the cooperative through the first few weeks of the new cocoa buying season. The cooperatives had to arrange for additional financing through 90-day notes with the Ghana Commercial Bank.

Many of the old-time cooperative members resented the government's action, viewing them as heavy-handed interference with the movement. Veteran committee members only grudgingly acknowledged the improved performance in cooperative operations. In 1974, some cooperative unions and societies showed a surplus, and paid their first bonus in four years. Also, some unions and societies were allowed to grant short-term advances to selected farmers. In 1975, the GCMA management committee granted local autonomy to selected cooperative unions and societies. From all appearances, cooperatives were on their way to establishing a strong financial footing.

Then, between 1975 and 1977, events were to overwhelm the cooperatives. During that two-year period, and especially during the 1976-77 season, it is widely believed that millions of cedis earmarked for cocoa purchases by the cooperatives were embezzled by officials in the Cocoa Marketing Board and top government organizations. The 1975-76 cocoa buying season was in many ways a repeat of the 1968 season. Scattered unions and societies experienced sporadic shortfalls in cash for purchasing cocoa. CMB advances to the cooperatives were lower than requested. Cocoa began to pile up as unions and societies awaited delivery of cash to purchase produce from the farmers and to pay for transporting cocoa from the countryside.

The 1976-77 purchasing season brought full crisis. Cocoa advances in many cases met only 20 percent of requirements. Meanwhile, with record world prices contrasting with the dwindling value of the cedi and falling real local cocoa prices, smuggling reached epidemic proportions. In 1976-77 an estimated 50,000 tons of cocoa, representing about one-sixth of the total harvest, appears to have been smuggled to the Ivory Coast and Togo. Corruption began to flourish again in cooperative operations.

Fraud in the cooperatives, however, was a trivial phenomenon when compared to the colossal embezzlement by senior government officials. Many believe the Acheampong government abolished the cooperatives in April of 1977 as a way of hiding rampant corruption in the top ranks of the state. Many sources argued that the cooperatives were simply convenient scapegoats for far more serious malfeasance by top officials.

The day after the cooperatives were abolished, editorials in all the leading Ghanaian papers praised the government action. Blame for elimination of the cooperatives was squarely placed on the cooperatives themselves. Even the discredited corpse of the United Ghana Farmers Cooperative Council was praised. After the second death of the cooperatives, there seemed no hope for a new resurrection.

The unhappy experience of cooperation reborn in Ghana again demonstrates the crucial importance of effective performance, and the impact of government support. In contrast to Uganda, the Ghanaian cooperatives have never since independence benefited from sustained, patient, supportive government action. In the 1960s and early 1970s, returns to cocoa farming were much less than they had been in the 1950s; it was accordingly much more difficult for the movement to generate rural savings on its own hook. When the government provided only episodic and uncertain backing, the obstacles were simply too great to be overcome. Forced into chronic improvisation, the cooperatives could no longer provide the reliable and attractive marketing services upon which they had built their strength in the 1950s. As a consequence, they lost the following they once had; long gone is the enthusiasm for rebuilding cooperation which surfaced with the fall of Nkrumah in 1966.

In both Ghana and Uganda, new regimes in 1979 tried to begin the now gargantuan task of repairing countries devastated by misrule. In Uganda, policy-makers assume as a matter of course that the cooperative machinery will be retained and refurbished in the cotton and coffee sectors. In Ghana, such an alternative no longer figures in the policy calculus.

12. Farmers, Cooperatives, and the Cocoa Marketing System in Ghana

The preceding two chapters dealt with the development and breakdown of the Ghanaian cooperatives. We saw the importance of farmer control and participation for the growth and development of the cooperatives, and the role of larger farmer-traders in organizing the cooperatives. Services and benefits were crucial in attracting farmers to the cooperatives. However, after 1966 cooperatives never regained their earlier effectiveness, in good part because of the failure of the government to offer sustained and consistent support.

With this complex and checkered history, what appraisal did Ghanaian farmers have of their cooperatives? Were members wholly disillusioned by the post-1966 shortcomings of cooperatives? We conducted a questionnaire survey in 1973-74 of farmers who were members and nonmembers of cooperatives as a complement to our other interviews and documentary research, to see what insight might be gained on farmer attitudes toward the cocoa cooperatives.

We interviewed 720 farmers during 1973 and 1974.[1] The farmers lived in all six cocoa growing regions of the country. Six hundred and sixty were members of cooperatives. The remaining 60 marketed through the Produce Buying Agency (PBA), set up as a subsidiary of the parastatal Cocoa Marketing Board in 1966. The PBA, in essence, inherited the assets and facilities of the liquidated UGFCC. The GCMA and the Department of Cooperatives alerted the cooperatives and helped find and notify farmers of the interviews. The PBA also cooperated but less so, as the dis-

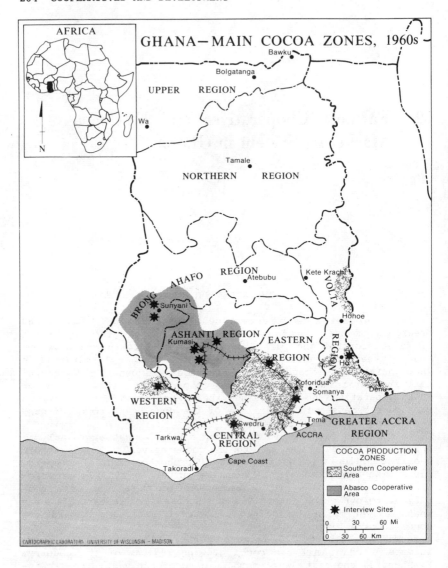

crepancy in the sample size suggests. Nonetheless, we needed
only a small control group and the PBA group satisfied the
requirement.

We stratified our sample according to the regional break-
downs in the cocoa farming population. We further stratified the
cooperative sample by union and society. At the society level,
we selected the farmers randomly. The PBA members were drawn
from the same village as cooperative members and were, when possible, neighbors of cooperators.

SOCIAL COMPOSITION OF THE COOPERATIVES AND THE PBA

The CPP accused the cooperatives of representing the interests of the larger, wealthier cocoa farmers. We wanted to verify
this claim, to the extent that subjective self-classification
permits. We assumed that if the cooperatives had a relatively
larger number of members who believed themselves to be in higher
social strata than the Produce Buying Agency clientele, the CPP
perception had some merit.

The survey data on farmer income did not show a majority,
or even a sizable plurality, of cooperative members in a single
social stratum (see Table 12.1). Indeed, only a small minority

Table 12.1

SUBJECTIVE ECONOMIC STATUS
(expressed as a percentage of respondents)

| | STATUS DESIGNATION | | | | |
Affiliation	Much Better Off	Above Average	Average	Below Average	Much Below Average
Cooperatives: Northern[a]					
(N = 398)	11	22	38	19	8
PBA: Northern					
(N = 19)	5	16	37	21	21
Cooperatives: Southern[b]					
(N = 217)	14	32	26	20	8
PBA: Southern					
(N = 32)	22	6	25	41	6

[a] Ashanti, Brong Ahafo.
[b] Volta, Eastern, Cape coast, some of Western regions.

viewed themselves as really well off, and two-thirds felt themselves of average means or less. However, the cooperative members did subjectively define themselves somewhat higher than PBA affiliates.[2]   But cooperators by 1974 did not see themselves as a "kulak" class.

Questions on farmers income level produced somewhat similar results (though accuracy of response is obviously uncertain). Over half of both cooperative and PBA affiliates claimed an income of less than 500 cedis. Only 7 percent of the cooperators (though 11 percent of PBA affiliates) said they had earned over 2,000 cedis.

Thus, taken as a whole, the cooperatives were not simply a club for the rich, in aggregate terms. But the relatively wealthy minority did control the leadership. In society after society, the more prosperous and influential farmers (and sometimes local chiefs) were the elected officers.

However, we should note that in 1973-74, the cooperative membership had swollen from pre-1961 levels, which could explain the large number of poor farmers in the membership. Before 1961, membership stood at 50,000. In 1973-74, some 200,000 cocoa farmers were affiliated to the cooperatives.

The widespread image of cooperatives as a consortium of wealthy farmers, then, derived above all from the nature of the leadership. Also, two-thirds of the cooperative members were Ashanti and Brong farmers, whose incomes (as recorded by both survey responses and other evidence) were higher than those of their counterparts in southern Ghana. Ashanti and Brong farmer predominance in the cooperatives helped foster the impression that cooperation represented wealth.

Table 12.2

AGE OF COCOA TREES, COOPERATIVE AND PBA MEMBERS
(expressed as a percentage of respondents
for each category of affiliation)

Affiliation	AGE OF COCOA TREES		
	5-15 Years	15-30 Years	Over 31 Years
Cooperatives: Northern (N = 376)	31	52	17
PBA: Northern (N = 20)	40	45	15
Cooperatives: Southern (N = 216)	19	33	47
PBA: Southern (N = 33)	21	52	27

Cooperative social composition might have changed radically had cooperatives been permitted to continue operating after 1977. Our data show that the cooperatives were not very successful in attracting younger farmers with higher future income potential. The PBA had more farmers with younger trees, and therefore higher potential for future income per acre than the cooperative farmers (Table 12.2). Further, somewhat more PBA farmers had larger landholdings, giving them more potential to expand planting if they chose. Had the competitive marketing system continued to operate beyond 1977, PBA might have become more of a "prosperous" farmer organization, and the cooperatives less so.

However, all cocoa farmers—cooperative and PBA—show a relatively high proportion of older cultivators. The contrast with the Uganda sample is quite striking (Table 12.3), and underlines the well-known weakness of Ghana's cocoa industry: its maturity. The farmers and their trees were older and less productive with each passing year.

Table 12.3

AGE OF GHANA COCOA FARMERS AND
UGANDA COTTON AND COFFEE FARMERS SURVEYED
(percentages)

Age	Ghana (1973-74) (N = 655)*	Uganda (1966-67) (N = 466)
Over 50	59.1	27.7
40-50	24.6	23.4
30-40	12.8	28.5
20-30	3.3	19.7
Under 20	0.2	0.7

* Cooperative members only.

COOPERATIVE SERVICES AND BENEFITS

Member participation and support is obviously critical to successful cooperative development; this in turn, we have argued, depends on reliable marketing and other services and material benefits. Ghanaian cooperative officials and experts echoed this theme, arguing that membership support rose and fell with avail-

able services and benefits. According to them, the dramatic decline in the level of cooperative services and benefits after
1966 led to erosion in member involvement and support.

To further explore this issue, we asked the farmers questions about participation, services, and benefits. Did farmers
join the cooperatives primarily to get services and benefits?
Or did they join for other reasons? Did they receive the benefits they expected. Did their success or failure in getting benefits influence their loyalty and support? The responses convinced us that farmer support and participation were built on one
central foundation: a desire for better financial returns, services, and benefits (Table 12.4). When asked if they had received the benefits they expected, a majority of the cooperative
farmers said they had. Just under one-third said they had not
(Table 12.5); overall these responses were similar to those in
Uganda. The time frame is important here: at the moment of our
Ghana interviews, cooperative performance had temporarily improved, under the impetus of military-imposed reform.

Pressed for examples of services provided by the cooperatives, the farmers gave the responses summarized in Table 12.6.
Cooperative staff members often said that bonuses were the most
common reason for joining, but only 27 percent of the respondents
mentioned them. When asked if they had received a bonus for the
previous season, 56 percent of the farmers replied affirmatively.
However, the bonus was hardly satisfying at the time the interviews were conducted. For the 1973-74 cocoa season, each farmer
received only 3 cedis for every 36 bags sold to the cooperative.
At this rate of return, a farmer with an annual income of 250 ce-

Table 12.4

REASONS FOR JOINING COOPERATIVE
(N = 655)

Reasons for Joining	Percentage of Cooperative Members
Bonus	12
Bonus and loan	15
Quick payment	11
Buy commodities	1
Inputs	12
Support cooperatives	26
Peer and family	8
Location near	4
No answer	12

Table 12.5

SATISFACTION WITH COOPERATIVE SERVICES
(percentages)

Has your cooperative given you the things you expected when you joined?	Ghana (N = 655)	Uganda (N = 269)
Yes	56.9	47.6
Some	5.3	21.6
No	30.2	24.0
Don't know; no response	7.5	6.8

Table 12.6

SERVICES RECEIVED BY MEMBERS
(N = 655)

Services Received	Percentage of Cooperative Members
Bonuses and loans	27
Technical assistance (i.e., advice)	16
Cash advances	13
Quick payment for cocoa	8
Commodities	4
No response (i.e., those who responded negatively to the receive benefits question)	33

dis from cocoa received a 2 cedi bonus. As this sum purchased only two yams at a rural market in 1973-74, the bonus payment was hardly enough to generate farmer enthusiasm.

The cooperatives provided other services which, we believed, might influence member support. From 1966 on, the cooperatives sold farm implements, commodities, and sprays on a limited basis. Beginning in 1972, the NRC administration asked the cooperatives (and PBA) to take over some of the distribution of inputs from private traders, thus increasing their volume considerably. Both the cooperatives and PBA appear to have reached many of the members. As Table 12.7 shows, slightly more than half of the farmers had received some type of commodity or farm input.

Table 12.7

DELIVERY OF SERVICES, COOPERATIVES AND PBA*
(percentages)

Services	YES Cooperative	PBA	NO Cooperative	PBA	DON'T KNOW NO RESPONSE Cooperative	PBA
Implements	57	58	33	35	10	7
Commodities	58	65	36	32	7	3
Sprays	37	25	55	70	9	5
Advice	56	40	36	57	8	3

* For cooperatives, N = 653; for PBA, N = 60.

Other services often mentioned by farmers were short-term cash advances on future purchases, short-term loans (which are indistinguishable from advances), and savings facilities. But these services were not available after the 1966 reorganization. Many cooperative staff members and officers attributed decline in member participation to the disappearance of these once-popular services.

We expected that, based on our interview with cooperative officials, small bonuses and the end of once-popular services would have weakened member participation, but our survey evidence did not confirm this presumption. By and large, members felt they were participating actively in the cooperatives; as Table 12.8 shows, the level of participation was quite high absolutely, and roughly comparable to Uganda. Even farmers who ranked on the low end of a scale of cooperative benefits received continued to participate.

This discrepancy between our expectations and findings can perhaps be explained by the qualitative difference in pre-1961 cooperative participation, when there were far more activities and more member involvement was required in overseeing and managing services and funds. Since these services were no longer available, member participation was by now limited to attendance at the annual meeting.

Table 12.8

MEMBER PARTICIPATION, GHANA AND UGANDA COOPERATIVES
(percentages)

Frequency of Attendance at Society Meetings	Ghana (N = 655)	Uganda (N = 256)
Always	56.6	31.2
Usually	10.6	26.9
Sometimes	20.3	34.0
Never	12.3	7.8
N.A.	2.0	

THE PBA COMPARED

The PBA did not require farmers to participate or otherwise demonstrate support for the organization. It was strictly a commercial operation, and, from our interviews, we concluded that PBA farmers liked it that way. They expressed solid support for their organization.

Farmers joined the PBA with no expectations of services and benefits other than a commercial marketing outlet. The PBA did little to change these expectations, at least until 1973. (In 1973, the NRC-appointed director of the PBA ordered the organization to distribute surpluses to the members as bonuses, thus departing from the PBA AND UGFCC policy of no bonuses.[3]) Our survey data revealed that for 50 percent (N = 30) of the PBA sample, the only reason for "joining" the PBA was to market their cocoa. For the remaining 50 percent (N = 30), reasons for supporting the PBA were:

- quick payment, 23 percent (N = 14);
- a nearby shed, 12 percent (N = 7);
- membership of father/family/friends, 7 percent (N = 4);
- government ownership, 5 percent (N = 3).

None of the responses suggest that the PBA farmers expected special financial benefits or other services when they joined the organization. However, the PBA bonus payment undoubtedly won support from the farmers. The PBA had not only promised but delivered—and had delivered more than it promised. After only one season of bonus payments, more than half the PBA farmers received a bonus.

Our conclusion, while it may seem obvious, is that a marketing organization cannot be successful unless commercial considerations are allowed to dominate management decisions. Had the
Ghanaian cooperatives concentrated more on making their organizations commercially effective--with or without short-term benefits
and services to the members--the cooperative movement might have
had a greater chance of survival. The PBA performance and growing array of successful services, at the time of our survey, impressed many farmers.

A COMPARISON OF COOPERATIVE AND PBA MANAGERIAL PERFORMANCE

The NRC administration established a tightly controlled,
competitive cocoa marketing system. Beginning in late 1972, only
the cooperatives and the PBA were permitted to purchase cocoa for
resale to the Cocoa Marketing Board. Both the cooperatives and
the PBA began the 1972 cocoa season with huge debts incurred over
the previous several years, making efficient crop purchase impossible. Recognizing this, the NRC wrote off their debts to the
Cocoa Marketing Board and ordered the CMB to make new financing
available. As a result, the PBA and the cooperatives began the
1972 cocoa buying season on an equal and relatively strong footing. Were they able to manage their organizations efficiently
as a result?
We wanted to compare the performance of the two marketing
organizations on the basis of farmer perceptions. Both organizations were charged with mismanagement and corruption during the
1968-71 period. The Department of Cooperatives and the Ghana Cooperative Marketing Association claimed that the NRC's strong
anti-corruption measures and reorganization dramatically curbed
corruption and reduced mismanagement. Did the farmers also share
this view?
The survey data showed that many farmers were aware of the
massive corruption and mismanagement that affected the cooperatives before the NRC takeover of the government. The responses
also suggest that farmers believed the NRC reorganization and
anti-corruption campaign to be moderately effective in curbing
corruption and mismanagement. Many farmer members began to take
a positive view of cooperative management and honesty, and ap-
plauded the government's efforts to curb corruption.
When asked, "Is there corruption and mismanagement in the
cooperative movement?", 29 percent of the farmers (N = 191) responded yes, 45 percent (N = 296) responded no, and 26 percent
(N = 168), "Don't know/no answer." When asked, "Is there less
corruption in the cooperative movement now than three years
ago?", 49 percent of the farmers responded yes (N = 319), 19 percent no (N = 127), and 34 percent (N = 209) that they didn't know

(or didn't answer). These responses suggest that the cooperatives were to some degree reducing corruption and mismanagement,
but had by no means eliminated it. Over a quarter of the farmers
believed these problems still existed. (The fact that so many
farmers appeared unaware of corruption or did not respond to the
question raises the question as to whether the pervasiveness of
corruption had been exaggerated.) When asked if the government
was "doing enough to stop corruption in the cooperative movement," the farmers responded: yes, 50 percent (N = 309); no,
6 percent (N = 36); don't know/no answer, 36 percent (N = 230).

Overall, the farmers felt that the cooperative management
was performing better than it had in the past, judging from responses to the question, "Do you think your cooperative is working (better, same, worse) than three years ago?" The responses
were: better, 67 percent (N = 440); same, 17 percent (N = 111);
worse, 12 percent (N = 77); don't know/no answer, 4 percent (N =
27). Presumably, the government's reorganization of the cooperatives and the infusion of new financing had had positive results.

Cooperative management received surprisingly high marks from
farmer members on honesty. When asked if the cooperatives were
honest in their business dealings with the farmers all the time,
sometimes, or never, the farmers responded:

Always	Mostly	Sometimes	Never	Don't Know/No Answer
68%	19%	8%	2%	4%
(N = 442)	(N = 121)	(N = 50)	(N = 13)	(N = 29)

This further suggests that the government-imposed reforms had
been effective.

PBA farmers gave their marketing organization high marks for
management performance. On the question of the PBA'S honesty,
the member farmers responded:

Always	Mostly	Sometimes	Never	Don't Know/No Answer
80%	8%	3%	3%	8%
(N = 48)	(N = 5)	(N = 2)	(N = 2)	(N = 5)

The PBA farmers consistently rated their organization's performance higher than the cooperatives. When asked if the PBA is
"more, as, less efficient in its business than the cooperatives,"
the response was:

- more, 65% (N = 39);
- as, 20% (N = 12);
- less, 3% (N = 2).

Sixty-five percent of the PBA members regarded their organization as "more" honest than the cooperatives.

CONCLUSION

The evidence obtained from our 1973-74 survey of cocoa
farmers supports some of the themes common to discussions of the
growth and development of cooperatives in Ghana.   The farmers
were attracted to the cooperatives largely because of the services and benefits they offered.   When those services and benefits were threatened, the members began withdrawing their support.   Yet there was still an important reservoir of member good
will, reflected in the quite positive assessments offered on a
number of dimensions of appraisal.   This in turn partly mirrored
the   short-lived   improvement   in   cooperative   functioning   in
1972-73.   But our interviews did reveal a renewed disappointment
with cooperative performance; this presaged the declining loyalty
in the later 1970s leading up to abolition in 1977.

13. Summary and Conclusions

In Chapter 2 we set forth a framework for the analysis of cooperatives' role as an instrument of rural and agricultural development in third world nations. Cooperatives were characterized as formally democratic organizations. Membership in them obliges farmers to yield some degree of autonomy in the management of their agricultural operations to the benefit of the cooperative.

Many of the advantages often attributed to cooperation are, in fact, advantages of large-scale formal organization. Thus, we proceeded to ask why farmers, and national political leadership, choose to act through cooperatives rather than alternative organizational forms. Furthermore, we noted that cooperatives constitute an institution mediating between farmers and government; thus we raised the issues of cooperative autonomy versus governmental control. Having identified motives which may produce a preference for cooperatives, we sought to suggest the implications of farmer-government interaction within the cooperative framework for the realization of the development goals of efficiency and equality.

In the rural communities of third world nations, a process of redefinition and crystallization of group identities is underway, spurred by increasing contact and conflict with broader regional and national societies. In this situation, small farmers are drawn to cooperation as the preferred means of improving their lot when they can be inspired to see in membership an expression of group solidarity in the face of some antithetical

element. Rural elites are drawn to the leadership of cooperatives where cooperative principles enhance their ability to mobilize followers' capital and enthusiasm for undertakings serving
their own values and ambitions, whether narrow, selfish designs,
or principled, community service objectives.

GOVERNMENT AIMS IN COOPERATIVES

Government attitudes and policies typically overshadow autonomous motivations. Where government chooses to encourage cooperation, its policy measures tend to become the dominant element pointing farmers toward cooperation. We do not discount the
possibility that elements of national political leadership may
be drawn to cooperation by their attachment to the principles of
cooperative ideology, usually as part of a more broadly socialist
commitment. However, more practically and more demonstrably, national political leadership is drawn to cooperation as a half-way
house between direct state implementation of economic activities
and state intervention limited to the regulation of private
enterprise.

Cooperation is chosen when, on the one hand, the institutionalized privileges of private property set unacceptably con-
fining limits to regulatory activity, and, on the other hand,
direct state operations entail too great a strain upon available
resources and too great a risk of failure. Part of the burden
of resource mobilization and responsibility for failure of economic programs may be shifted from the state to cooperatives.
Yet as a relatively new institutional form, dependent upon government backing, cooperatives are both legally and practically
susceptible to a far higher degree of governmental regulation and
intervention than private economic undertakings.

Autonomy versus Control

Government policy determines certain basic parameters which
constitute centrally important elements in the operational environment of cooperative organizations. However, detailed direction of the operations of local level societies is typically beyond the means of national political leaders and the administrative bodies responsible for the enforcement of cooperative policy. Intensive intervention at the local level is therefore limited to ad hoc attempts to deal with crisis situations. Secondary and tertiary cooperative organizations (unions and unions of
unions), by virtue of their smaller number and greater economic
power, attract more intense governmental attention. Political
leadership's inclination to restrict cooperative autonomy and
exercise more direct and detailed control will be a function of
two main factors: cooperatives' ability to fulfill their economic

responsibilities, and the political acceptability of cooperative leadership.

We have seen that in both Ghana and Uganda the impetus for cooperative organization in the early, pre-independence period grew out of a sense of economic grievance which was tied to a growing ethnic and national awareness. Thus, when farmers roused themselves and leaders sought to mobilize them to action, many chose not the limited company, but rather the cooperative, as a means of expressing and formalizing their sense of shared interests and shared identity. Leaders found that they could achieve both political and economic ambitions through the cooperatives, using political and social sentiments to help win backing for their economic initiatives.

In Uganda, in particular, the nationalist political tinge of cooperation was strong. Indeed, in the largest, most spectacular of the cooperative organizations—Musazi's Uganda African Farmers Union—the political aspect clearly dominated, while the basis of economic activity was weak. In Ghana, by contrast, though the connection between marketing arrangements and policy inevitably gave political significance to cooperatives' economic endeavors, overt political activism of cooperative organizations was less visible. Indeed, the movement came to be seen by the populist CPP leadership as a conservative cartel of capitalist farmers.

An explanation for this difference may be found in the differing political economy of the main export crops in Ghana and Uganda. In Uganda, the in-country marketing of cotton and coffee was inextricably linked with preliminary processing activities which required relatively large investments in processing facilities. In this situation, non-African interests devoted considerable efforts to the elaboration of a network of governmental regulation aimed at protecting their investments in processing. Together, high entry capital requirements and regulatory restrictions severely limited African involvement in the in-country handling of export crops. Thus, there was little room for African cooperative involvement in cotton or coffee without government measures aimed at removing legal impediments and providing the economic means required for the acquisition and operation of ginneries and coffee processing factories. From the early days of Ugandan cooperation, then, cooperative activity was necessarily accompanied by strong political overtones.

In Ghana, by contrast, capital requirements for the in-country handling of cocoa were minimal and few legal impediments were raised to African involvement in in-country trade. Issues such as the cocoa export cartel's 1937 attempt to limit the price paid to African traders could arouse the cooperatives and the African agricultural and trading community in general to joint actions with racial-political overtones; in general, however, a more intensive involvement in economic activities was open to Ghanaian cocoa cooperatives without the necessity of waging political campaigns to achieve government assistance.

In the post-independence period, governmental policy came to overshadow spontaneous popular enthusiasm as the main factor drawing farmers to membership in Ugandan cooperatives. The political implications of cooperation were clear to the UPC leadership which guided Uganda to independence, and generally positive ties to the leadership of the cooperative movement led politicians to believe that the political import of cooperation could be exploited to their benefit. While one of the major cooperative unions was located in Buganda, whose ethnic sub-nationalism was the UPC's major political opponent, the cooperative movement was not dominated by the Ganda. Thus, the UPC chose cooperatives as a major instrument for the displacement of non-African economic interests in the agricultural sector; for the extension of governmental control and influence over the agricultural economy; and for the promotion of agricultural innovations.

Within this framework of support for the cooperative mechanism, the UPC government found it necessary to apply an intensifying level of discipline in its relations with the cooperatives in the period from 1967 onward. Most dramatically, the elected leaders of a number of the most important cooperative unions were stripped of their authority and management of the unions was placed in the hands of supervising managers appointed by the Cooperative Department. The need to control and intervene did not alienate the UPC government from the cooperatives as a device for development and control of the economy, however. To the contrary, cooperative domination of the major export crops was strengthened by policy measures introduced during this period.

This finding is not surprising when considered against the background of our analysis of national political leadership's motives for promoting cooperation. If the unions could function independently in a competent and politically reliable manner, so much the better. If not, UPC leaders had explicitly justified favoritism toward cooperatives in terms of government's ability to intervene more easily and more radically in cooperative affairs, whenever necessary. While supervising managers conducted the union's affairs, the network of local societies continued to function without imposing a direct burden on government administrative capabilities. The societies could be adequately directed and overseen from the level of the union, as the restoration of economic balance in unions placed under supervisors seems to indicate. Furthermore, it should be noted that toward the end of the UPC regime, supervising managers were withdrawn from a number of unions and the formal autonomy of the unions was restored.

By the time Amin had seized power, cooperatives were a basic fixture of the rural economy and an important institutional tool standing at government's disposal. Even an impulsive, erratic dictator, suspicious of cooperative leaders' political loyalty, did not consider deliberately harming such a valuable instrument.

The expansion of cooperatives' monopsonistic control over increasing sectors of agricultural marketing was frozen, but the basic cooperative structure was left intact. Cooperatives suffered, of course, from the devastation of Uganda wrought by Amin's regime, but they were not subjected to measures intended to damage them in particular as cooperatives.

In post-independence Ghana, government policy in general held negative implications for cooperatives and their members. Indeed, it is striking that the successes of cooperatives in the period leading up to the establishment of the CPP regime left an impression upon officers and members strong enough to produce a revival of cooperatives after a period of outright repression. The post-independence government's hostility to cooperatives derived, first of all, from the fact that the cooperative movement was centered in the anti-Nkrumah, Ashanti cocoa-growing areas. Those cooperative leaders who were politically active were associated with the CPP's opponents. In addition to the ethnic tensions which alienated the CPP and Nkrumah from the Ashanti, the regime's hostility to local Ghanaian smalland medium-scale entrepreneurs, such as the prosperous cocoa planters dominating the cooperative movement, also added to the determination to undermine the cooperatives. Thus, Nkrumah first bypassed and then destroyed the cooperatives, replacing them with alternative organizational forms.

The declared economic policy orientation of Nkrumah's immediate successors, the NLC and PP governments, seemed to indicate a more liberal approach within which cooperatives might be restored to their former position in the cocoa trade. In fact, however, cooperation was only tolerated. Cooperatives did not receive the positive backing which would have enabled them to overcome the damage caused by the confiscation of their assets by the Nkrumah regime. Two factors accounted for the cooperatives' inability to win positive, active government support. First, the new leadership was under pressure from private trading interests, from the parastatal Cocoa Marketing Board bureaucracy, and apparently from the government-owned Ghana Commercial Bank as well, not to favor cooperatives. Thus, the government gave only ambiguous and episodic support to the revived cooperatives, and never returned the bulk of their confiscated assets. By the time of the NRC's rise to power in 1972, cooperative affairs were in disarray. The government first imposed a heavy tutelage, which did produce some short-lived improvement in performance. The effectiveness of state support soon declined, and cooperatives were overwhelmed in the general institutional decay. They were formally liquidated in 1977.

Development Goals: Equality

What is to be expected concerning cooperatives' ability to contribute to the realization of the development goals of equal-

ity and efficiency, given the motives of national political leadership, and typical small farmers, as they have been depicted
above? On the equality issue, it has been our contention that
the values and structure of village society are such as to miti-
gate against the realization of equality through the autonomous
activity of local cooperative societies. The abstract concept
of human equality is foreign to village thought, and, on the
whole, the vertical ties of dependence upon powerful patrons tend
to outweigh horizontal ties among equals for the members of the
lower strata of village society. Those apparently leveling devices found in the village are typically aimed at nothing more
than the partial relief of the most difficult situations. Moreover, such remedies are intended only for the limited circle of
individuals meeting the local definition of community membership.
Given these characteristics, it is clearly unreasonable to expect
that in the normal course of events internal village forces will
mobilize to resist local elites' manipulation of cooperatives so
as to reinforce their already advantaged position.

An egalitarian dynamic will develop within cooperative
structures, then, only when some supra-village force is at work
which possesses both the will and the means to overcome the influence of local society's stratification structure. The application of outside resources makes possible the implementation of
measures essential to the realization of egalitarian objectives.
First of all, the character of cooperative leadership would have
to be altered. Existing cooperative leadership, drawn from local
elites, might be pressured or tempted to act in a manner more
congenial to development goals. Alternatively, existing leadership may be replaced by new cadres, recruited from the lower
strata of village society. At least initially, such new leadership would be dependent upon outside backing against the compet-
ing veteran elites. Thus, the new leaders might be expected to
be far more susceptible to the guidance of outside elements than
their predecessors, who possess an autonomous, local power base.
In addition, outside intervention might be aimed directly at encouraging rank-and-file members to greater awareness and activity
within the cooperatives, so that membership pressure would come
to serve as a check upon leadership abuses.

Elements of the national political leadership may be inclined to act in such a manner as to promote equality for reasons
of ideological persuasion. However, practical steps would be
possible only where the national leadership was strong enough and
willing to risk, if not to provoke openly, confrontation with local elites. The pragmatic return for such a willingness to enter
into conflict might be found in the prospect of strengthening
ties to and influence over both the leadership and the mass of
the village population.

Neither the Ghanaian nor the Ugandan cases, nor African cooperative experiences more generally, offer much support for the

prospect of such a policy being systematically pursued, or suc-
ceeding even if it were attempted. Such a high degree of state
intervention at the local level presumes both an extraordinary
competency on the part of the bureaucratic apparatus, and also,
importantly, an exceptional capacity for public-regarding action.
We have seen, in both the Ghanaian and Ugandan instances, that
government capacity to really supervise and control the local
agencies of cooperation—the primary societies—was quite circumscribed; intervention was mainly centered on the regional cooperative unions serving a number of primary societies. Both countries had, in the 1960s, relatively competent and effective public services, whose performance in this regard might be said to
define the outer limits of the possible in Africa more generally.
    There are also other exemplary cases to be noted, where re-
gimes with a strong egalitarian ideological commitment at the
center sought for a time to make cooperation a key component in
redistributive rural strategies. Tanzania pursued the policy
most consistently over the longest time period; yet, in 1975 and
1976, the Nyerere regime was led to abolish both primary and secondary cooperatives. The reason advanced for abandonment of such
a cherished experiment was the inability of the state to prevent
cooperatives from becoming an autonomous institutional base for
the most prosperous and respected farmers, whose policy preferences were often antithetical to the rural socialism preached by
Nyerere. Hyden has aptly summarized this situation:

> There is little doubt . . . that the closure of all
> co-operatives was also determined by the fact that the
> bureaucratic bourgeoisie were unable effectively to con-
> trol the co-operative organizations. The leadership of
> the co-operative movement often had divergent views from
> party and government officials on matters relating to ru-
> ral development. By virtue of their position as leaders
> of organizations with a local power base, they could ar-
> gue effectively with party and government leaders. More-
> over, they could mitigate the pressures on the peasants
> by the party . . . . Because they controlled a large
> share of the exchanges between government and peasants
> but were not fully incorporated into the party and gov-
> ernment machineries, they were making policy implementa-
> tion more difficult.[1]

    Beyond the issue of state capacity, there is the question
as to whether attempted state intervention at this level might
not be predominantly shaped, in practice, by the higher aim of
fostering political support for the regime, or the simple bureaucratic goal of asserting hegemony over the village periphery in
the interests of the state (or, as some would argue, of the administrative bourgeoisie whose social interests are most closely

tied to the state).[2]   This was, we have been in Ghana, clearly
the net effect of the CPP "cooperative" instrument, the UGFCC.
In Zambia, where cooperatives were briefly promoted as an agency
of the Kenneth Kaunda version of humanistic socialism, leadership
and benefits went especially to the rural faithful who had served
as local organizers in the nationalist campaign of the United National Independence Party (UNIP).[3]   Similar outcomes have been
noted in Guinea and Mali, where cooperation had a time in the sun
as institutional expression of a populist socialist orientation.[4]

Development Goals: Efficiency

    As for the possibility of achieving efficiency objectives
through cooperation, we have noted above that local and national
political leadership are each faced with a choice with regard to
management of cooperative affairs. On the one hand, it is possible to strive to achieve efficiency in cooperative functioning,
and thus to provide relatively marginal benefits to the many. On
the other hand, toleration of unequal benefits to the more successful farmers, or even misappropriation, makes it possible to
offer substantial benefits to a small minority. The latter
course of action, of course, offers more easily realizable short-
run political gains.
    The inclination to reap immediate benefits at the expense
of efficiency may be moderated at both local and national levels
by ideological commitment and ideals of service. More practically, where alternative marketing outlets or sources of services
are available, local and national leadership will have to take
into account the possibility that the plunder of cooperative resources will quickly bring the total liquidation of cooperative
activity. For while the dependent status of the rank-and-file
members neutralizes their ability to raise demands for efficiency
and to back such demands by the application of political pressures within the cooperative, farmers are in many instances sufficiently independent to "vote with their feet," leaving the cooperative an empty shell.
    More positively, national political leadership may be drawn
to intervention aimed at guaranteeing efficient cooperative operations by the possibility of achieving political returns in the
medium to long run. As with regard to equality, effective intervention offers the prospect of realizing greater popular backing
if it does help ensure that the most valued benefits of coopera-
tion--reliable and remunerative marketing services--are provided.
We believe that greater participation by members is in fact induced by successful performance of these economic tasks. Local
leaders may find a positive motive for efficient management where
they themselves have a significant, farm-based economic interest
in their cooperative's ability to provide efficient services on
a long-run basis.

We conclude that not only Ghanaian and Ugandan experience, but accumulated evidence on cooperation elsewhere in Africa unmistakably demonstrate that efficient performance of assigned economic functions is the sine qua non of cooperative viability.[5] Sustained, patient state support is indispensable to attaining reasonable levels of efficiency.[6] A gamut of degrees of state intervention is observable in achieving these ends. In Egypt, state control and domination are pervasive, consistent with an historically enrooted statist tradition;[7] in Ghana, the cooperatives in their halcyon years in the 1950s were quite autonomous. We should also note that intervention is likely to founder if it goes beyond support and tutelage into imposition of forms of cooperation which large numbers of farmers regard as running counter to their material interests.[8]

Equality and Efficiency: The Ugandan and Ghanaian Experience

We found in both Uganda and Ghana that autonomous local forces at work within cooperative structures tended to reinforce existing inequalities within African communities. We have noted domination of office and the distribution of special benefits, such as cooperative credit, by members of the dominant strata of rural society. Little effective governmental intervention was undertaken in either Uganda or Ghana in order to counteract these tendencies.

To the extent that control of cocoa cooperatives by rural capitalists perturbed the Nkrumah government, it responded by disbanding the cooperative movement rather than attempting to struggle within it against anti-egalitarian trends. The cooperatives were too strong to be changed from within; it seemed to the CPP more reasonable to liquidate them, and work through alternative organizational structures. The cure, we have seen, proved worse than the disease. For reasons already discussed, later Ghanaian governments showed only ambivalent interest in promotion of the reborn cooperatives, or in a refined and difficult intervention in cooperative affairs so as to enable them to serve as a policy instrument for the accomplishment of complex development objectives.

In Uganda, the radical, populist rhetoric of the UPC's later years found little practical expression in conscious, systematic attempts to promote equality through cooperatives. The radical leadership elements in the UPC were not strong enough, nor was the UPC party apparatus sufficiently effective and autonomous, to make possible a drive to achieve egalitarian objectives through cooperatives. Intervention came only in order to restore financial balance and minimal standards of efficiency and honesty in instances where inefficiency and exploitation threatened to bring the total collapse of important cooperative unions. Indeed, the populistic rhetoric of the UPC party leadership hardly penetrated the relevant administrative agency, the Cooperative Department, and had even less impact upon actual field operations. Amin, of course, could not serve as the source of inspiration for measures aimed at introducing more egalitarian standards into the activity of Ugandan cooperatives.

On the other hand, it would be wrong to draw solely negative conclusions with regard to the implications of cooperation for equality. However much the wealthy and powerful may have siphoned off a disproportionate share of cooperative resources, the system of cooperative bonuses provided at least a partially effective means of allowing farmers to share in the profits to be made from processing and trading in agricultural produce. Though mismanagement often prevented important segments of the cooperative movement from paying bonuses, the failure was never absolute. In Uganda, the belief in the producers' right to share in such profits came to be rooted in the consciousness of the Ugandan farming populace, as a result of cooperative activity. Similarly, while the wealthy and powerful took more than their share of cooperative credit, a share still remained for the rank and file. The competition for cooperative loans was in any case far more open than in other governmentally sponsored credit programs which, given the definition of their target public ("progressive farmers") or production branch (beef ranching, dairy farming), automatically eliminated average farmers from the pool of eligible candidates.

In short, as a group-oriented, formally egalitarian framework, the cooperative symbolized, and to a modest degree embodied, an alternative to either the purely statist marketing and service-providing structures or simple reliance on private markets. For certain development programs, the cooperatives came to constitute a possible means of implementation, relatively more committed and better suited to the realization of egalitarian objectives. While far from an ideal realization of cooperation's potential in this field, Ugandan cooperatives are an institutional asset which might be exploited by a suitably oriented national political leadership.

As with regard to equality, our review of the efficiency of cooperative operations reveals a mixed performance. Ghanaian cooperatives prospered in a competitive environment in their early period, a reliable sign of efficient functioning; and, indeed, it is reasonable to attribute the cooperatives' efficiency to their need to stand against the pressures of competition. Cooperative difficulties in the period following their restoration were no doubt aggravated by internal management weaknesses, but the basic problem would seem to have arisen from the terms under which the cooperatives were allowed to resume operating. Without compensation for the assets Nkrumah had confiscated and without adequate provision of credit for investment and operating expenses, the cooperatives faced a nearly impossible task in their attempts to reclaim their share of the market.

More generally, one cannot help being struck by the paradoxical frequent correlation between successful cooperation and rural capitalism. Cooperation, we argued in our opening chapter, historically arose as a reaction to capitalism, and an alternative to it. Yet, especially in Ghana, the cooperative movement was to an important degree the expression of petty capitalism in the cocoa areas. A similar observation might be made for the relatively effective cooperative movement in Kenya. In Tanzania, the cooperative movement in the 1950s experienced its most spectacular triumphs in the export agriculture zones, where commercial farming on petty capitalist premises was most widespread.[9]

In Uganda, an important decline in the efficiency of cooperative operations was registered in the years following independence. The new African government provided the cooperative movement with massive backing and thus promoted rapid growth in both the number of cooperatives and the extent of their operations. This rapid expansion produced openings for abuse and inefficiency which would seem to be partly the result of the diffusion and dilution of government's supervisory activities. In addition, there developed a feeling of immunity against Cooperative Department intervention as a result of the rise to power of politicians closely tied to the cooperative leadership.

By the time the UPC had completed the political maneuvers which eliminated the KY and left the DP impotent, the losses of important cooperative unions had reached such dimensions as to oblige harsh corrective measures. The new political situation made such measures possible; and, as noted above, the UPC remedial programs had a significant effect in ameliorating cooperative performance. While local cooperatives continued to function in a more or less independent fashion, the unions placed under the control of Cooperative Department officials were returned to financial balance, and in some cases at least to formal autonomy.

Moreover, it should be noted that the UPC not only clamped down upon elected cooperative leadership. The political leadership itself retreated from a number of politically inspired decisions which had been forced upon the Cooperative Department and which had contributed significantly to the deterioration of the cooperative credit scheme's recovery rate. The overrapid expansion of the cooperative credit program was halted and a moderate pace of growth came to prevail. The group farm loans diminished to insignificant sums, as the project's failure became clearly evident. On the other hand, the rapid expansion of tobacco loans granted in Acholi and Lango during the period under discussion constituted a politically inspired burden upon the credit scheme, counterbalancing the gains achieved as a result of the decisions just mentioned.

In sum, the Ugandan experience indicates that the inefficiencies and mismanagement of the early post-independence period's overrapid, forced growth were not an irreparable problem. After a period of excessive freedom from governmental oversight, a period of intensive intervention succeeded. Once it produced the necessary changes, the restoration of a more normal situation of checks and balances among governmental supervision, cooperative leadership, and cooperative membership could be begun. Indeed, as the support-intervention/displacement-support sequence in the dairy sector illustrates, elements within the governmental apparatus came to realize, in some instances at least, that direct administration of development activities was a less efficient approach than cooperative operations.

CONCLUSIONS

In our study of cooperative operations in Uganda and Ghana we have noted two major factors limiting the developmental potential of cooperatives: first, the weakness of the typical small farmers, those most in need of the advantages which cooperation could theoretically bring; and, second, the pressures of economic and political self-interest which are felt by leaders at the local and national levels with the greatest power to shape the character of cooperative operations. While reviewing these difficulties we have at the same time tried to stress two factors which moderate the severity of the doubts which our analysis arouses. First, we have pointed to the special attraction which the democratic membership principle of cooperation may hold where farmers have developed a clear sense of group identity in opposition to other socioeconomic groups. This basic, defining feature of cooperation can in such situations play an important role in crystallizing farmer solidarity into positive, productive, joint efforts.

Secondly, and perhaps more significantly, we have emphasized the relative standards of judgment which must be applied to evaluation of cooperation as a policy instrument. State intervention is a basic fact of economic life in the underdeveloped countries, as are politically insecure regimes, resource limitations, and limited and overstrained policy-making and administrative capabilities. Thus, doubts and criticism concerning the efficacy of cooperatives must be tempered by the realization that: (a) some form of government intervention is unavoidable; and (b) governments experiencing problems in their promotion of cooperation are likely to encounter similarly intractable difficulties, for similar reasons, in attempts to realize their goals through alternative policy devices. Moreover, although cooperation's promise of broad-based democratic involvement of the common man in processes of social and economic change is difficult of realization, the alternative means of intervention in the economy--regulation of private undertakings and direct governmental conduct of agricultural operations--seem to offer even more limited prospects of allowing the mass of small farmers an active role in the transformation of their societies.

Notes

NOTES TO CHAPTER 1

1. Louis P.F. Smith, The Evolution of Agricultural Co-operation (Oxford: Basil Blackwell, 1961), p. 4.
2. G.N. Ostergard and A.H. Halsey, Power in Cooperatives
(Oxford: Basil Blackwell, 1965).
3. Guy Hunter, "Agricultural Administration and Institutions," Conference on Strategies of Agricultural Development in
the 1970s, Food Research Institute, Stanford University, December
1971.
4. Goren Hyden, Beyond Ujamaa (Berkeley: University of California Press, 1980).
5. Robert Bates, States and Markets (Berkeley: University
of California Press, forthcoming).
6. Margaret Digby, The World Cooperative Movement, rev. ed.
(London: Hutchison University Library, 1960).
7. Smith, Agricultural Co-operation, p. 60.
8. Digby, World Cooperative Movement, p. 8.
9. But not always; the Hutterites in the prairies of Canada
and the United States have maintained close-knit, economically
successful cooperative farming units for generations. Peter Dor-
ner, ed., Cooperative and Commune (Madison: University of Wisconsin Press, 1977), pp. 65–90.
10. Digby, World Cooperative Movement, pp. 21–50.
11. Ostergard and Halsey, Power in Cooperatives, p. 199. The
authors note a declining vitality of the participative structures, whose members are now mainly elderly, and of ideological
commitment, except from a minority of activists, many of whom are
also employees.

12.   John W. Bennett, "Agricultural Cooperatives in the Development Process: Perspectives from Social Sciences," California Agricultural Policy Seminar Monograph no. 4 (University of California-Davis, May 1979), p. 3, cites evidence revealing very diffuse ideological understandings of cooperatives. In one Arkansas survey, almost half the members could not cite a specific responsibility of a member, and half were unsure who "owned" the cooperative (p. 31).

13.   Digby, World Cooperative Movement, pp. 74-86; Daniel Thorner, Agricultural Cooperatives in India: A Field Report (Bombay: Asian Publishing House, 1964), pp. 25-26.

14.   Konrad Engelman, Building Cooperative Movements in Developing Countries (New York: Praeger, 1968), pp. 10-11.

15.   Hans Dieter Seibel and Andreas Massing, Traditional Organizations and Economic Development: Studies of Indigenous Co-operatives in Liberia (New York: Praeger, 1974).   Bennett, "Agricultural Cooperatives," lays stress upon this point.

NOTES TO CHAPTER 2

1.   Throughout this chapter, the terms "cooperation" and "cooperative" should be understood as referring to agricultural cooperation and cooperatives.

2.   See, for example, Margaret Digby, The World Cooperative Movement, rev. ed. (London: Hutchinson University Library, 1960).

3.   For some representative studies see: John S. Saul, "Marketing Cooperatives in a Developing Country," in P. Worsley, ed., Two Blades of Grass (Manchester: Manchester University Press, 1971), pp. 347-70; Carl Gosta Widstrand, ed., Cooperatives and Rural Development in East Africa (Uppsala: Scandinavian Institute of African Studies, 1970); E.J. Schumacher, Politics, Bureaucracy and Rural Development in Senegal (Berkeley: University of California Press, 1975); Franz Schurmann, Ideology and Organization in Communist China (Berkeley: University of California Press, 1968), Chapter VII; Joan Vincent, "Rural Competition and the Cooperative Monopoly: A Ugandan Case Study," in J. Nash, J. Dandler, and N. Hopkins, eds., Popular Participation in Social Change: Cooperatives, Collectives and Nationalized Industry (The Hague: Mouton, 1976); Goran Hyden, Efficiency Versus Distribution in East African Cooperatives (Nairobi: East African Literature Bureau, 1973); Raymond Apthorpe, ed., Rural Cooperatives and Planned Change in Africa: An Analytical Overview (Geneva: United Nations Research Institute for Social Development, 1970); S. Bunker, The Uses and Abuses of Power in a Uganda Farmers' Marketing Association: The Bugisu Cooperative Union, Ltd. (Ph.D. diss., Duke University, 1975).

4.   See Dudley Seers, "The Meaning of Development," in Norman T. Uphoff and Warren F. Ilchman, eds., The Political Economy

<u>of</u> <u>Development</u> (Berkeley: University of California Press, 1972),
pp. 123-26; see also Hyden, <u>Efficiency Versus Distribution</u>.

5.   In practice, in the best-known form of collective encountered in the Soviet Union, the theoretical importance of membership is so circumscribed by the predominance of managers who are
directly subordinated to state directives and control that the
correspondence of this form with the cooperative concept virtually disappears.   In theory, however, the collective may be seen
as a limiting case of cooperative farm organization.

6.   John W. Bennett, "Agricultural Cooperatives in the Development Process: Perspectives from Social Science," California Agricultural Policy Seminar Monograph no. 4 (Davis: University of
California, May 1979), p. 5.   We are indebted to Bennett for
bringing this point to our attention.   For a valuable study of
indigenous forms of cooperation in a contemporary setting, see
Hans Dieter Seibel and Andreas Massing, <u>Traditional Organizations</u>
<u>and Economic Development: Studies of Indigenous Co-operatives in</u>
<u>Liberia</u> (New York: Praeger, 1974).

7.   T. Lowi, <u>The End of Liberalism</u> (New York: W.W. Norton,
1969), pp. 3-41.

8.   B.F. Johnston, "Agricultural and Economic Development:
The Relevance of the Japanese Experience," <u>Food Research Insti-</u>
<u>tute Studies</u>, 6, no. 3 (1966), p. 255.

9.   John M. Brewster, "Traditional Social Structures as Bar-
riers to Change," in Herman M. Southworth and Bruce F. Johnston,
eds., <u>Agricultural Development and Economic Growth</u> (Ithaca, N.Y.:
Cornell University Press, 1967), pp. 92-94.

10.   Samuel P. Huntington, <u>Political Order in Changing Soci-</u>
<u>eties</u> (New Haven, Conn.: Yale University Press, 1968), Chapter 1.

11.   For a general statement of the need to evaluate policy
in the underdeveloped countries in terms of the weakness of central governments, see Joel S. Migdal, "Policy and Power: A Framework for the Study of Comparative Policy Contexts in Third World
Countries," <u>Public Policy</u>, 25, no. 2 (Spring 1977), pp. 241-60.

12.   For Ghana, see the discussion below in Chapters 10-12.
For Uganda, see Department of Co-operative Development, <u>Annual</u>
<u>Report for the Year Ended 31st December, 1966</u>; <u>1967</u>; <u>1968</u>; <u>1969</u>;
<u>1970</u>; <u>1971</u>, Mimeo. (Kampala: Department of Co-operative Develop-
ment).   During 1971, less than a third of the active agricultural
cooperative unions were under the direction of a government-
appointed supervising manager; the supervised unions accounted
for some 55-60 percent of the total turnover of agricultural
unions in the 1970/71 crop year.   (Lack of detail and internal
inconsistencies in Departmental reports make presentation of more
precise figures impossible.)

13.   In a number of African states, many farm households are
headed by women, either because of separation from husband by
death or divorce, or more frequently because the husband holds
wage employment in town.   In addition, women play a critical role

in the division of farm household labor, even when the husband
is present. See Kathleen A. Staudt, Agricultural Policy, Political Power, and Women Farmers in Western Kenya (Ph.D. diss., University of Wisconsin-Madison, 1976). "Himself" here and in subsequent usage should be understood as meaning "him or herself."

14. Wyn F. Owen, "The Double Developmental Squeeze on Agriculture," American Economic Review, 16, no. 1 (March 1966),
pp. 49-50; Jeffrey M. Paige, Agrarian Revolution (New York: Free
Press, 1975).

15. For a description of an extreme case of small farmer
persistence, see Clifford Geertz, Agricultural Involution (Berkeley: University of California Press, 1968); see also Colin Leys,
"Politics in Kenya: The Development of Peasant Society," British
Journal of Political Science, 1, no. 3 (July 1971), pp. 307-38.
Of course, many do desert the farm in order to attempt to improve
their lot in the urban areas.

16. Owen, "Double Developmental Squeeze," pp. 54-56.

17. Meir Merhav, Technological Dependence, Monopoly, and
Growth (Oxford: Pergamon Press, 1969), pp. 16-65.

18. See, for example, E.A. Brett, Colonialism and Underdevelopment in East Africa (New York: Nok Publishers, 1973),
pp. 237-65.

19. An indication of the extent to which prejudice and the
inability to comprehend the true causes of economic distress may
be operative in creating the negative image of the trader may be
found in Vernon W. Ruttan, "Agricultural Product and Factor Markets in Southeast Asia," in K.R. Anschel, Russell H. Brannon, and
E.D. Smith, eds., Agricultural Cooperatives and Markets in Developing Countries (New York: Praeger, 1969), pp. 79-106.

20. See, for example, M. Lewin, Russian Peasants and Soviet
Power (New York: W.W. Norton, 1975), Chapter 1.

21. An interesting case study in which many of the points
made in the following analysis may be noted is K. Galla, Sociol-
ogy of the Cooperative Movement in the Czechoslovak Village
(Prague: Country Life Association, 1936), Chapter V.

22. See Peter B. Clark and James Q. Wilson, "Incentive Systems: A Theory of Organizations," Administrative Science Quarterly, 6, no. 1 (September 1961), pp. 130, 134-36; and James Q.
Wilson, Political Organizations (New York: Basic Books, 1973).

23. This is analogous to the peasant-landlord relationship
in the insightful analysis of James C. Scott, The Moral Economy
of the Peasant (New Haven, Conn.: Yale University Press, 1976).

24. See Guy Hunter, Modernizing Peasant Societies (New York:
Oxford University Press, 1969), pp. 42-48; Joel Migdal, Peasants,
Politics and Revolution (Princeton, N.J.: Princeton University
Press, 1974), Part Three; Keith Griffin, The Political Economy
of Agrarian Change (London: Macmillan, 1974), pp. 229-30.

25. See Scott, Moral Economy; Migdal, Peasants, Politics and
Revolution, Chapters II and III.

26. Ibid., Part One.

27. For critical evaluation of such assumptions see Earl M. Kulp, Rural Development Planning (New York: Praeger, 1970), pp. 60-62; D.K. Leonard, Reaching the Peasant Farmer (Chicago: University of Chicago Press, 1977), pp. 177-86.

28. See H.U.E. Thoden Van Velzen, "Staff, Kulaks and Peasants: A Study of a Political Field," in L. Cliffe et al., eds., Government and Rural Development in East Africa (The Hague: Martinus Nijhoff, 1977), pp. 223-50.

29. See Willard W. Cochrane, Agricultural Development Planning (New York: Praeger, 1974), pp. 10-11.

30. These are the advantages as depicted in theory; in practice many problems tend to arise. See the discussion in Kulp, Rural Development Planning, Chapter 23; D.M. Hunt, "The Ugandan Agricultural Co-operative Credit Scheme," East African Journal of Rural Development, 5, nos. 1 and 2 (1972), pp. 1-38; and D.M. Hunt, Credit for Agricultural Development (Nairobi: East African Publishing House, 1975).

31. Griffin, Political Economy, pp. 221-22.

32. See, for example, Geoff Lamb, Peasant Politics (London: Julian Friedmann Publishers, 1974); Joan Vincent, An African Elite (New York: Columbia University Press, 1971); G.C. Hickey, Village in Viet Nam (New Haven, Conn.: Yale University Press, 1964), Chapter 9.

33. For discussions of the benefits of cooperative office, see: Vincent, An African Elite; Stephen G. Bunker, "Strategies for Upward Mobility in Bugisu District, Uganda," typescript (1971); E.A. Brett, "Problems of Cooperative Development in Uganda," in Apthorpe, Rural Cooperatives and Planned Change, pp. 127, 135; Presidential Special Committee, Report of the Presidential Special Committee of Enquiry into Cooperative Movement and Marketing Boards (Dar-es-Salaam: Government Printer, 1966); Committee of Inquiry, The Report of the Committee of Inquiry into the Affairs of the Busoga Growers Cooperative Union Limited (Entebbe: Government Printer, 1965); Committee of Inquiry, The Report of the Committee Inquiry into the Affairs of All Cooperative Unions in Uganda (Entebbe: Government Printer, 1968).

34. See Mancur Olson, The Logic of Collective Action (New York: Schocken, 1971), Chapters I and II.

35. James S. Coleman, "The Development Syndrome: Differentiation-Equality-Capacity," in L. Binder, et al., Crises and Sequences in Political Development (Princeton, N.J.: Princeton University Press, 1971), pp. 73-100.

36. See I. Kopytoff, "Socialism and Traditional African Societies," in William H. Friedland and Carl G. Rosberg, eds., African Socialism (Stanford, Calif.: Stanford University Press, 1964), pp. 53-62; Lewin, Russian Peasants, Chapter 1; Migdal, Peasants, Politics and Revolution, pp. 66-82; Marc Bloch, French Rural History (Berkeley: University of California Press, 1966),

pp. 45-48; Goran Hyden, Beyond Ujamaa in Tanzania (Berkeley: University of California Press), pp. 18-19, 113-14.

37. The experience of Communist movements in Southeast Asia is instructive here. One notes the stress placed on nationalist themes of xenophobic unity as the major principle of village level organization in the earlier period of Communist uprisings, and the disastrous consequences of a sharpening of intra-village tensions before achievement of adequate political power in the Indonesian case. See Chalmers Johnson, Peasant Nationalism and Communist Power (Stanford, Calif.: Stanford University Press, 1962); Douglas Pike, Viet Cong (Cambridge: MIT Press, 1966), p. 41; Joseph Buttinger, Viet Nam: Dragon Embattled (New York: Praeger, 1967), pp. 206, 213; Eric Wolfe, Peasant Wars of the Twentieth Century (New York: Harper and Row, 1969), pp. 184-85; Donald Hindley, The Communist Party of Indonesia 1951-1963 (Berkeley: University of California Press, 1964), pp. 171-77; Guy Pauker, "Political Consequences of Rural Development Programs in Indonesia," Pacific Affairs, 41, no. 3 (Fall 1968), pp. 338-91; Guy Pauker, The Rise and Fall of the Communist Party of Indonesia (Santa Monica: Rand Corporation Memorandum RM 5753-PR, 1969); Rex Mortimer, Indonesian Communism under Sukarno: Ideology and Politics 1959-1965 (Ithaca, N.Y.: Cornell University Press, 1974).

38. Uganda Government, Government White Paper on the Report of the Committee of Inquiry into the Coffee Industry, 1967 (Entebbe: Government Printer, 1968), p. 4.

39. See Tony Killick, Development Economics in Action (New York: St. Martin's Press, 1978), pp. 21-26.

40. Uganda National Assembly, Parliamentary Debates, Vol. 102 (July 1970), pp. 72-73; "A New Deal for Farmers," Uganda Argus, 14 February 1970, p. 1; "Action on Cost of Living," ibid., 1 March 1971, p. 1; S.D. Ryan, "Economic Nationalism and Socialism in Uganda," Journal of Commonwealth Political Studies, 11, no. 2 (July 1973), p. 153.

41. See, for example, Colin Leys' discussion of the African "bourgeoisie" in Underdevelopment in Kenya (Berkeley: University of California Press, 1975), pp. 51, 165-69, 174-78.

42. See statements by M. Ngobi (from the right) and J. Kakonge (left) in Uganda National Assembly, Parliamentary Debates, Vol. 10 (March-April 1963), pp. 425, 437 (Ngobi); and Vol. 84 (July 1968), pp. 3690, 3693 (Kakonge).

43. The particular difficulties of agricultural policy-making and administration are examined by Cochrane, Agricultural Development Planning; Griffin, Political Economy, Chapters 8, 10, 12, 13, 14; Kulp, Rural Development Planning; B.F. Johnston and P. Kilby, Agriculture and Structural Transformation (New York: Oxford University Press, 1975); Leonard, Reaching the Peasant Farmer; David Leonard, ed., Rural Administration in Kenya (Nairobi: East African Literature Bureau, 1973); Uma Lele, The Design of Rural Development: Lessons from Africa (Baltimore: Johns Hop-

kins University Press, 1975); R. Chambers, Managing Rural Development: Ideas and Experience from East Africa (Uppsala: Scandinavian Institute of African Studies, 1974); Jon R. Moris, The
Agrarian Revolution in Central Kenya: A Study of Farm Innovation
in Embu District (Ph.D. diss., Northwestern University, 1970);
Guy Hunter, The Administration of Agricultural Development (London: Oxford University Press, 1970).

44. The absence of strong, independent external bodies moni-
toring the operations of governmental administrative units is
stressed in Fred Riggs' influential, pioneering analyses of development administration; see, for example, "Bureaucrats and Political Development: A Paradoxical View," in Joseph La Palombara,
ed., Bureaucracy and Political Development (Princeton, N.J.:
Princeton University Press, 1963), pp. 120-67; Fred Riggs, Administration in Developing Countries (Boston: Houghton Mifflin,
1964). See also G.D. Ness, Bureaucracy and Rural Development in
Malaysia (Berkeley: University of California, 1967), pp. 12-21.

45. J.B.B. Isabirye, The Duties and Functions of the Department of Co-operative Development, Mimeo. (Kampala: Department of
Co-operative Development, 1973), Paragraph 42.

NOTES TO CHAPTER 3

1. Among the more important works are David Apter, The Po-
litical Kingdom in Uganda, 2nd ed. (Princeton, N.J.: Princeton
University Press, 1967); Nelson Kasfir, The Shrinking Political
Arena (Berkeley: University of California Press, 1976); Mahmood
Mamdani, Politics and Class Formation in Uganda (New York:
Monthly Review Press, 1976); D.W. Low, Buganda in Modern History
(Berkeley: University of California Press, 1971); D. Anthony Low
and R. Cranford Pratt, Buganda and British Overrule (London: Ox-
ford University Press, 1960); Cherry Gertzel, Party and Locality
in Northern Uganda, 1945-1962 (London: Athlone Press, 1974); F.G.
Welbourn, Religion and Politics in Uganda, 1952-1962 (Nairobi:
East African Publishing House, 1965); D. Anthony Low and Alison
Smith, History of East Africa, vol. 3 (Oxford: Clarendon Press,
1976), Low and Gertzel contributions; Donald Rothchild and
Michael Roben, in Gwendolen M. Carter, ed., National Unity and
Regionalism in Eight African States (Ithaca, N.Y.: Cornell Uni-
versity Press, 1966); James H. Mittelman, Ideology and Politics
in Uganda (Ithaca, N.Y.: Cornell University Press, 1975); Ali
Mazrui, Soldiers and Kinsmen in Uganda (Beverly Hills, Calif.:
Sage Publications, 1975).

2. For a detailed analysis of the crucial 1900 agreement,
see Low and Pratt, Buganda and British Overrule, pp. 1-159.

3. C.C. Wrigley, Crops and Wealth in Uganda (Kampala: East
African Institute of Social Research, 1959), p. 16. See also the
same author's contribution on the Uganda economy, 1903-1945, in

Vincent Harlow and E.M. Chilver, eds., <u>History of East Africa</u>, vol. 2 (Oxford: Clarendon Press, 1965), pp. 395-475.

4.   Ibid., pp. 400-406.

5.   Low and Pratt, <u>Buganda and British Overrule</u>, p. 237.

6.   Henry W. West, <u>Land Policy in Buganda</u> (Cambridge: University Press, 1972), pp. 70-78.

7.   Wrigley, <u>Crops and Wealth</u>, pp. 27-40.

8.   Lucy Mair, <u>An African People in the Twentieth Century</u> (London: George Routledge and Sons, 1934), p. 278.

9.   E.S. Hayden, "The History of the Bugisu Coffee Scheme," Ms. (Mbale, Uganda, 1953).

10.   Fallers, <u>Bantu Bureaucracy</u> (Chicago: University of Chicago Press, 1965), p. 149.

11.   Mamdani, <u>Politics and Class Formation</u>, p. 104.   This study offers an excellent account of the interwar period, pp. 65-119.   See also E.A. Brett, <u>Colonialism and Underdevelopment in East Africa</u> (London: Heinemann, 1973).

12.   Mamdani, <u>Politics and Class Formation</u>, p. 106.

13.   The best coverage of party development in these years remains Low, <u>Buganda in Modern History</u>, pp. 167-226.

14.   Though many Ugandans thus identified the UPC as a Protestant party, this was only partly true.   Donald Rothchild and Michael Rogen, in Carter, ed., <u>National Unity and Regionalism</u>, pp. 379-84, argue on the basis of 1962 election results that by this time at least the party was no longer distinctly Protestant.

15.   There is some variation in these estimates.   J.J. Oloya, <u>Coffee, Cotton, Sisal and Tea in the East African Economies</u> (Nairobi: East African Literature Bureau, 1969), advances figures of 100 man-days per acre for coffee, and 122 for cotton.   Dennis Lury, in Low and Smith, eds., <u>History of East Africa</u>, vol. 3, p. 219, gives 80 and 140, respectively.   Yield calculations are rendered somewhat problematic by the discovery in an agricultural census conducted in 1963-1965 that the administration had substantially over-estimated crop acreages.   Thus the 1962 Cotton Commission gives as the average yield per acre for 1952-1962, 292 pounds/acre, and Lury uses the figure of 500.

16.   G.B. Masefield, <u>Agricultural Change in Uganda 1945-1960</u> (Stanford, Calif.: Food Research Institute, 1962), p. 123.   The levy was introduced in 1948 at 1 shilling and raised to 1.50 in 1950.   The colonial authorities claimed this local government bonus led to doubled output in 1949 over 1948.

17.   Joan Vincent, "Rural Competition and the Cooperative Monopoly: A Ugandan Case Study," in J. Nash, J. Dandler, and N. Hopkins, eds., <u>Popular Participation in Social Change: Cooperatives, Collectives and Nationalized Industry</u> (The Hague: Mouton, 1976), pp. 70-97.   The colonial administration, of course, did not suppress other crops, but never identified one it felt worth promoting in the cotton zones.

18.   E.H. Winter, <u>Bwamba Economy</u> (Kampala: East African Institute of Social Research, 1955), pp. 34-35.

19. Vali Jamal, "Taxation and Inequality in Uganda, 1900–1964," Journal of Economic History, 38, no. 2 (June 1978), pp. 418–38.

20. R.M.A. Van Zwanenberg and Anne King, An Economic History of Kenya and Uganda 1800–1970 (London: Macmillan Press, 1975), p. 203.

21. Uganda Government, Report of the Uganda Cotton Commission, 1938 (Entebbe: Government Printer, 1939), pp. 87–88.

22. Uganda Government, Report of the Uganda Cotton Industry Commission, 1948 (Entebbe: Government Printer, 1948), pp. 13–14. It is interesting to compare this figure with the estimates of 14–20 percent peasant loss through fraudulent buying in the pre-war Nigerian oil palm industry; here the middlemen were Nigerians. Eno J. Usoro, The Nigerian Oil Palm Industry (Ibadan: Ibadan University Press, 1974), p. 57.

23. Uganda Cotton Commission, 1948, p. 5.

24. Winter, Bwamba Economy, p. 35.

25. The reaffirmation of central hierarchical control is analyzed by S. Griffith, Local Politicians and National Policies: The Secretaries-General of Uganda (B. Phil. thesis, University of Oxford, 1969). The vitality of the elected district governments during their heyday in the 1950s and 1960s is well portrayed in Colin Leys, Politicians and Policies (Nairobi: East African Publishing House, 1967); see also District Government and Politics in Uganda (Madison: University of Wisconsin African Studies Program, 1978).

26. Irvin Gershenberg, "Slouching Toward Socialism: Obote's Uganda," African Studies Review, 15, no. 1 (April 1972), pp. 79–85.

27. The World Bank team noted that careful study of foreign trade figures showed "a substantial outflow of private capital from Uganda that must have taken place for a number of years." International Bank for Reconstruction and Development, The Economic Development of Uganda (Baltimore: Johns Hopkins University Press, 1962), p. 30.

28. Neal Sherman, A Political-Economic Analysis of Ugandan Dairy Policy (Ph.D. diss., University of Wisconsin-Madison, 1975), pp. 44–50.

29. J. Bayo Adekson, "Ethnicity, the Military, and Domination," Plural Societies, 9, no. 1 (Spring 1978), pp. 85–110. See also Holger Bernt Hansen, Ethnicity and Military Rule in Uganda (Uppsala: Scandinavian Institute of African Studies, 1977).

30. The most valuable sources on the Amin coup and regime are Samuel Decalo, Coups and Army Rule in Africa (New Haven, Conn.: Yale University Press, 1976); Mazrui, Soldiers and Kinsmen in Uganda; David Martin, General Amin (London: Faber and Faber, 1974); Aiden Southall, "General Amin and the Coup: Great Men or Historical Inevitability," Journal of Modern African Studies, 13, no. 1 (1978), pp. 85–106; Michael Lofchie, "Uganda Coup: Class Action by the Military," Journal of Modern African Studies, 10,

no. 1 (May 1972), pp. 19-36; International Commission of Jurists, Uganda and Human Rights (Geneva, 1977); David Gwynn, Idi Amin: Deathlight of Africa (Boston: Little Brown, 1977); Henry Kyemba, A State of Blood (New York: Ace Books, 1977).

31. Jack D. Parson, "Africanizing Trade in Uganda: The Final Solution," Africa Today, 20, no. 1 (1973), pp. 59-72; Michael Twaddle, ed., Expulsion of a Minority: Essays on Ugandan Asians (London: Athlone Press, 1975).

32. Various figures are given on the size of the Asian Community in 1972. Kyemba says the total was only 50,000, of whom 20,000 were citizens. State of Blood, p. 56.

33. Ibid., pp. 112-14.

34. Martin furnishes thorough documentation on this fiasco, and why it failed. Initially there was to have been a multi-pronged invasion from Sudan as well as Tanzania, but the Sudanese government had to remove the Obote camps as a part of the peace settlement with southern Sudanese insurgents, who had close ties to Amin. General Amin, pp. 170-97.

35. Lulakome A. Kayiira and Edward Kannyo, "Politics and Violence in Uganda," Africa Report, 23, no. 1 (January-February 1978), pp. 41-42.

36. Kyemba, State of Blood, pp. 51-52.

37. International Commission of Jurists, Uganda and Human Rights, p. 167.

38. Kyemba, State of Blood, p. 59.

39. African Research Bulletin, 14, no. 8 (15 September 1977), pp. 4334-35.

40. Uganda Government, Work for Progress: The Second Five-Year Plan 1966-1971 (Entebbe: Government Printer, 1966), p. 65.

41. Audrey Richards, Ford Sturroch, and Jean M. Fortt, Subsistence to Commercial Farming in Present-Day Buganda (Cambridge: Cambridge University Press, 1975), especially pp. 154-68.

42. Legislative Council, Debates, 30th Session, 2d Meeting (1951), p. 91.

43. G.K. Hellimen, "Economic Collapse and Rehabilitation in Uganda," Paper presented to African Studies Association annual meetings, Philadelphia, October 1980.

44. The most comprehensive assessment of the damage was by a team of experts recruited by the Commonwealth Secretariat, The Rehabilitation of the Economy of Uganda, 2 vols. (London: Commonwealth Secretariat, 1979).

45. Africa Contemporary Record, 1975-76, p. B344.

NOTES TO CHAPTER 4

1. See the analysis by Nelson Kasfir in Carl G. Widstrand, ed., Cooperatives and Rural Development in East Africa (Uppsala: Scandinavian Institute of African Studies, 1970), pp. 178-208.

2. Cyril Ehrlich, The Marketing of Cotton in Uganda 1900-1950 (Ph.D. diss., University of London, 1958), p. 222.
3. Ibid., p. 259.
4. Quoted in Uganda Government, Report on the Marketing of Native Produce (Entebbe: Government Printer, 1937), Appendix 4.
5. C.F. Strickland, "Cooperation for Africa," Africa, 6, no. 1 (January 1933), pp. 17-18.
6. Geoffrey F. Engholm, "The Decline of Immigrant Influence on the Uganda Administration 1945-1952," Mimeo. (Kampala: Makerere University, 1963[?]).
7. Ibid., p. 6.
8. Uganda Government, Commission of Inquiry into the Progress of the Co-operative Movement in Mengo, Masaka and Busoga Districts (Entebbe: Government Printer, 1952), p. 4.
9. Uganda Government, Report of the Committee of Inquiry into the Affairs of All Co-operative Unions in Uganda (Entebbe: Government Printer, 1968), p. 16.
10. The government policy of placing ginneries in cooperative hands proved to be a bonanza for Asian ginners. To begin with, the policy included elimination of "surplus capacity" by purchase of 35 silent ginneries, a 1 million pound windfall to owners of these redundant plants; these were purchased with funds from the Cotton Price Assistance Fund, constituted by holding down prices paid to growers. Further, at the same time as the legislation permitting compulsory acquisition of ginneries was adopted, the formula by which ginning costs were calculated was adjusted in a direction very generous to ginnery operators. One economic historian estimates the effective profit rate on ginning in the 1950s was 40 percent. (Lury, in D.A. Low and Alison Smith, History of East Africa, vol. 3 [Oxford: Clarendon Press, 1976], p. 232.) This immediately had the effect of inflating the market value of ginneries; by 1953, an Agriculture Department official noted the "puzzlingly high prices" at which ginneries were being sold. A number of ginnery operators succeeded in unloading their plants to cooperatives at exceptionally high prices, reflecting the state-created artificial market value. Others succeeded in bribing cooperative officials to use government concessional finance to purchase their often decrepit ginneries at premiumn prices. (E.A. Brett, "Problems of Cooperative Development in Uganda" [United National Research Institute for Social Development], p. 19.)
11. Report of the Committee of Inquiry into the Affairs of All Co-operative Unions in Uganda, p. 17.
12. Uganda Government, The Report of the Committee of Inquiry into the Affairs of the Busoga Growers Co-operative Union Limited (Entebbe: Government Printer, 1966), p. 14.
13. Report of the Committee of Inquiry into the Affairs of All Co-operative Unions in Uganda, pp. 368-69.
14. Affairs of the Busoga Growers Co-operative, pp. 8-9.

Stephen G. Bunker, "Forms and Functions of Government Intervention in a Uganda Cooperative Union," Paper presented to the African Studies Association annual meetings, Denver, Colorado, 1971, pp. 34-35.

15. Ibid.

16. Uganda Government, National Assembly, Debates, vol. 68 (1967), pp. 1781-89. Two years later Nekyon angrily informed Parliament that his own mother had been issued a worthless cooperative chit for her cotton.

17. Interviews with officials of the Agency for International Development and Agricultural Cooperative Development International, Washington, June 1972.

18. Commonwealth Secretariat, The Rehabilitation of the Uganda Economy, vol. 2 (London, 1979), p. 55.

19. Ibid.

20. Ibid., p. 56.

21. Department of Cooperative Development, West Acholi, Annual Report, 1964. On the issue of kondoism, Beverly Brock provides interesting evidence of popular concern with rural violence in documenting the emergence of neighborhood vigilante groups in the 1960s in many parts of Bugisu; "The Search for 'Social Constraints'," R.D.R. no. 87 (Kampala, Uganda: Makerere University, 1969). In Teso, a 1967 survey showed that kondos ranked above every other problem facing farmers; Victor C. Uchendu and K.R.M. Anthony, Agricultural Change in Teso District, Uganda (Nairobi: East African Literature Bureau, 1975), p. 91.

22. Joan Vincent, "Rural Competition and the Cooperative Monopoly," in J. Nash, J. Dandler, and N. Hopkins, eds., Popular Participation in Social Change: Cooperatives, Collectives and Nationalized Industry (The Hague: Mouton, 1976), pp. 70-97.

23. Ibid., pp. 93-97.

24. David Jacobson, Itinerant Townsmen: Friendship and Social Order in Urban Uganda (Menlo Park, Calif.: Cummings Publishing Company, 1973).

25. Report of the Committee of Inquiry into the Affairs of All Co-operative Unions in Uganda, p. 22.

26. Stephen Bunker, "Making It in Bugisu," Ms. (Durham, N.C., 1972[?]); The Uses and Abuses of Power in Uganda Farmers' Marketing Association: The Bugisu Cooperative Union, Ltd. (Ph.D. diss., Duke University, 1975).

27. Uganda, Legislative Council, Debates, 40th Session, 4th Meeting (1960), pp. 1995-2021.

28. Affairs of the Busoga Growers Co-operative Union, pp. 32-33.

29. Joel Barkan on Bunyoro, in District Government and Politics in Uganda (Madison: African Studies Program, University of Wisconsin, 1978).

30. A.B. Mukwaya, "The Rise of the Uganda African Farmers' Union in Buganda," Mimeo. (East African Institute of Social Research, n.d.).

31. Gambuze, 2 April 1948, quoted ibid., p. 4.

32. Stonehouse had a chequered subsequent career. He was a junior minister in the Wilson government in the early 1960s, then forced out of Parliament for fraudulent transactions and faking his own suicide.

33. John Stonehouse, Prohibited Immigrant (London: Bodley Head, 1960), p. 46. See also George W. Shepherd, They Wait in Darkness (New York: John Day, 1955).

34. Ibid., p. 93.

35. Ibid., pp. 234-49. See also Uganda, Report of the Ad Hoc Committee of Legislative Council appointed to consider and report upon the proposals for the Reorganisation of the Coffee Industry, The Coffee Industry Bill, and the Coffee (Export Duty) Bill (Entebbe: Government Printer, 1953), and Inquiry into the Progress of the Co-operative Movement in Mengo, Masaka and Busoga.

36. This process is given excellent formulation in Nelson Kasfir, The Shrinking Political Arena (Berkeley: University of California Press, 1976); and Holger Hansen, Ethnicity and Military Rule in Uganda (Uppsala: Scandinavian Institute of African Studies, 1977).

37. On the Gisu, see especially Joan La Fontaine, The Gisu of Uganda (London: International African Institute, 1959); for Sebei, see the comprehensive monograph by Walter Goldschmidt, Culture and Behavior of the Sebei (Berkeley: University of California Press, 1978). Local politics in Bugisu and Sebei are covered by Young in District Government and Politics in Uganda.

38. The early phases of cooperative development are ably analyzed by Bunker, Uses and Abuses of Power. See also E.S. Haydon, "The History of the Bugisu Coffee Scheme," Mimeo. (Mbale, Uganda, 1953).

39. Uganda, Report of the Commission of Inquiry into the Affairs of the Bugisu Co-operative Union Limited (Entebbe: Government Printer, 1958), p. 5.

40. The nature of the survey is described in Chapter 6.

41. The cooperatives, however, participated in the celebration of Buganda as an entity. Records of such Buganda-based unions as Uganda Growers contain many references to support for Kingdom government, gifts for the Kabaka on his birthday, and the like. We are endebted to Nelson Kasfir for this observation.

42. Uganda, Report of the Committee of Inquiry into the Cotton Industry 1966 (Entebbe: Government Printer, 1966); D.G.R. Belshaw, "Price and Marketing Policy for Uganda's Export Crops," East African Journal of Rural Development, 1, no. 2 (July 1968), pp. 33-49.

43. Biographical data are from Who's Who in East Africa (Nairobi: Marco Surveys, 1964).

44. Affairs of the Busoga Growers Co-operative Union, pp. 32-33.

45. Report of the Committee of Inquiry into the Affairs of All Co-operative Unions in Uganda, p. 30.

46. However, the Amin regime did subsequently accuse Obote of "politizing" the movement, as had the DP in earlier years. The 1971 Annual Report of the Cooperative Department makes this charge (¶143). While there is no doubt that the UPC would have liked to use the cooperative unions as a political auxiliary, and there was frequent political party factionalism at both union and society levels, in our judgment the predominant thrust of government policy was to use cooperatives as a state, not a party, instrument, and to generally uphold the old "apolitical" ideology of the Cooperative Department. In Bugisu, both DP and UPC were essentially congeries of personal factions, which helps explain the apparent anomaly of a DP Prime Minister supporting the BCMA while the BCU itself was headed by a DP man.

47. F.G. Welbourne, Religion and Politics in Uganda 1952-1962 (Nairobi: East African Publishing House, 1965).

48. Agricultural and Cooperative Development International, Progress through Cooperatives: Annual Report 1971 (Washington, D.C., 1971), pp. 14-19.

49. Inquiry into the Cotton Industry 1966, p. 41.

NOTES TO CHAPTER 5

1. Henry W. West, Land Policy in Buganda (Cambridge: University Press, 1972); Diana Hunt, Credit for Agricultural Development: A Case Study of Uganda (Nairobi: East African Publishing House, 1975), pp. 19-36.

2. Department of Agriculture, Annual Report of the Department of Agriculture for the Year ended 31st December, 1961 (Entebbe: Government Printer, 1962), p. 21.

3. Department of Cooperative Development, Annual Report . . . 1960 (Entebbe: Government Printer, 1961), p. 6; Annual Report . . . 1963, Mimeo. (Kampala: Department of Cooperative Development, 1964), p. 7.

4. Department of Agriculture, Annual Report . . . 1964 (Entebbe: Government Printer, 1968), p. 14.

5. Government of Uganda, Work for Progress--The Second Five-Year Plan 1966-1971 (Entebbe: Government Printer, 1966), p. 75.

6. Ibid., p. 65.

7. Joan Vincent, "Rural Competition and the Cooperative Monopoly," in J. Nash, J. Dandler, and N. Hopkins, eds., Popular Participation in Social Change: Cooperatives, Collectives, and Nationalized Industry (The Hague: Mouton, 1976), pp. 70-97.

8. Hunt, Credit for Agricultural Development, pp. 129-236.

9. Ibid., pp. 196-236.

10. Ibid., pp. 266-93.

11. D.C. Frederickson, Cooperative Specialist-Credit, "Novem-

ber 1971 Monthly Report," Ms., Kampala, American Cooperative Development International headquarters files, Washington, D.C.

12. Ministry of Animal Resources, Farmers' Forum Report 1971 (Entebbe: Government Printer, 1973); Ministry of Information and Broadcasting, The First 366 Days (Entebbe: Government Printer, 1972[?]).

13. Diana Hunt, "The Ugandan Agricultural Co-operative Credit Scheme," East African Journal of Rural Development, 5, nos. 1-2 (1972), pp. 36-37.

14. Hunt, Credit for Agricultural Development, pp. 129-236.

15. Hunt, "The Ugandan Agricultural Co-operative Credit Scheme," pp. 36-37. Guy Hunter, A.H. Bunting, and Anthony Bottral, Policy and Practice in Rural Development (London: Allanheld, Osmun & Co., 1976), argue that rural credit schemes for smallholders have often proved unsuccessful. They suggest that the capacity of rural mechanisms to generate savings without credit is greater than realized. For supporting evidence, see Hans Dieter Seibel and Andreas Massing, Traditional Organizations and Economic Development (New York: Praeger, 1974); Sara S. Berry, Cocoa, Custom, and Socio-Economic Change in Rural Western Nigeria (Oxford: Clarendon Press, 1975).

16. Commonwealth Secretariat, The Rehabilitation of the Economy of Uganda, 2 vols. (London: Commonwealth Secretariat, 1979), vol. II, p. 101.

17. William Allen, The African Husbandmen (Edinburgh: Oliver & Boyd, 1965), pp. 433-35.

18. J.B. Warren, "Co-operative Group Farming in Uganda," Conference of East African Machinery Specialists, Egerton College, Kenya, March 1965, p. 1; see also V.D. Jameson, Agriculture in Uganda, 2nd ed. (London: Oxford University Press, 1970), p. 199.

19. John Cleave and E.H. Jones, ibid., p. 119.

20. Work for Progress, p. 58.

21. Colin Leys, Politicians and Policies (Nairobi: East African Publishing House, 1967), p. 80.

22. Malcolm Hall, in Gerald Helleiner, ed., Agricultural Planning in East Africa (Nairobi: East African Publishing House, 1968), p. 115.

23. Ibid., p. 109.

24. Hunt, Credit for Agricultural Development, pp. 304-13; Department of Cooperative Development, Annual Report . . . 1969, p. 16.

25. "1969/70 Policy Speech to Parliament by Hon. J.B.T. Kakonge, M.P., Minister of Agriculture and Forestry," July 1969, pp. 15-16.

26. Uganda Government, Achievements of the Government of Uganda during the First Year of the Second Republic (Entebbe: Government Printer, 1972), p. 12.

27. Uganda Government, Uganda's Plan Three: Third Five-Year

Development Plan 1971/2-1975/6 (Entebbe: Government Printer,
1972), pp. 162-63.
    28. African Research Bulletin 13, no. 10 (30 November 1976),
p. 4071.
    29. Uganda's Plan Three, p. 167.
    30. Achievements of the Second Republic, p. 13.
    31. Commonwealth Secretariat, Uganda Economy; Donald Roth-
child and John Harbeson, "The Political Economy of Rehabilitation
in Uganda," Paper presented to African Studies Association annual
meetings, Philadelphia, October 1980.
    32. For representative figures, see National Assembly, De-
bates, vol. 100 (1970), p. 443.

                        NOTES TO CHAPTER 6

    1. The survey was conducted in selected parishes of eight
districts of Uganda in 1966 and 1967 by Young and E.A. Brett.
The districts chosen were Masaka and West Mengo (Buganda), Bugisu, Busoga, Bunyoro, Teso, Acholi, and Kigezi. The choice was
governed by the desire to include both predominantly coffee (Bugisu and Buganda) and cotton (Busoga, Teso, Acholi) unions. Bunyoro handled both coffee and cotton, as well as tobacco; Kigezi
had coffee and vegetables. In Buganda and Bugisu, the cooperative movement was relatively old; in Kigezi, it was very new.
In Bunyoro, Busoga, and Bugisu, cooperatives were very strong;
in Buganda and Kigezi, much less so. Within a district, several
parishes were chosen (two to six) which were, in the view of local government officers, reasonably representative. Within each
parish, a 5 percent random sample was drawn, based on the list
of taxpayers. The questionnaire, after pretesting, was administered in the local languages by Makerere University students.
It is important to bear in mind that, while random sampling was
used within parishes, the aggregated results do not constitute a
true national sample. We are grateful to Samuel Busulwa, James
Katerobo, Isaac Ojok, and Edward Takhuli for their assistance
with interviewing, and Marshall Carter, Kathleen Lockard, Catharine Newbury, and Thomas Turner for their labors in coding and
data analysis.
    2. Anthony Oberschall, "Communications, Information and As-
pirations in Rural Uganda," Journal of Asian and African Studies,
4, no. 1 (January 1969), pp. 30-50. Quite similar results for
Ghana are reported by Fred M. Hayward, "A Reassessment of Conventional Wisdom about the Informed Public: National Political Information in Ghana," American Political Science Review, 70, no. 2
(June 1976), pp. 433-51.
    3. The equation of "pagan" and poverty may relate to the
historical role of the Catholic and Protestant churches in oper-

ating rural schools and serving as point of contact with potential mobility channels. The "pagans," by definition, had never been to school (or they would have been required to convert).

4. Uganda Government, The Report of the Committee of Inquiry into the Affairs of the Busoga Growers Co-operative Union Limited (Entebbe: Government Printer, 1966); Uganda Government, Report of the Committee of Inquiry into the Affairs of All Co-operative Unions in Uganda (Entebbe: Government Printer, 1968).

5. Affairs of the Busoga Growers Co-operative Union, p. 27.

6. Victor C. Uchendu and K.R.M. Anthony, Agricultural Change in Teso District, Uganda (Nairobi: East African Literature Bureau, 1975), p. 54.

7. John W. Bennett, "Agricultural Cooperatives in the Development Process: Perspectives from Social Science," California Agricultural Policy Seminar Monograph no. 4 (Davis: University of California, May 1979), pp. 30-32.

8. Ibid., p. 18.

NOTES TO CHAPTER 7

1. Neal Sherman, A Political-Economic Analysis of Ugandan Dairy Policy (Ph.D. diss., University of Wisconsin-Madison, 1975), pp. 222-24.

2. For the purpose of this survey, a dairy farmer was defined as a farmer who had introduced exotic or crossbred stock into his herd; a farmer in the process of crossbreeding his local stock through use of the Veterinary Department's artificial insemination service; or a farmer, keeping only local cattle, who was a fully paid-up member of one of the three principal cooperative societies in Bulemezi, and who delivered milk to the society of which he was a member on a regular basis. Of the 142 farmers included in the survey, 123 fell into the first category, 11 into the second, and 8 into the third. The survey was carried out in 1973.

To the best of our knowledge the survey is virtually complete in its coverage of those meeting the definition given above. Cooperative officers were consulted as to identification and location of society members; Veterinary Department staff and inseminators were questioned; and the farmers themselves were asked to provide names of others also engaged in rearing stock of exotic blood. Only two of all those falling within our definition were unwilling to be interviewed, and thus are not included in the survey.

Thus, in essence, the survey constitutes a census of dairy farmers in the county; and we deal here with what should be viewed as a case study, rather than a sample survey. Statistical tests of significance are not particularly appropriate to the evaluation of such findings. Rather, the differences between groups and the association between certain variables as discussed in the text must be subjected to the same sort of informed judgment as employed in evaluating nonquantitative material derived from case studies.

The 142 dairy farmers form a small fraction of the no fewer than 5,000 farmers in Bulemezi who were engaged in raising cattle. This minimum estimate is based upon data presented in Donald S. Ferguson, An Economic Appraisal of Tick Borne Disease Control in Tropical Africa: The Case of Uganda (Ph.D. diss., Cornell University, 1971), p. 34; and B.W. Langlands, A Preliminary Review of Land Use in Uganda (Kampala: Department of Geography, Makerere University, 1971), p. 115.

3.   A number of cooperatives were officially gazetted as the sole agents of the Dairy Corporation in their areas; see Legal Notice no. 1 of 1968, Agents of the Dairy Corporation. Records of attempts to mobilize government backing to guarantee society control of local milk markets may be found in both government and society files. Among the East Buganda cooperatives which engaged in such lobbying were the following societies: Bugerere Balunzi, Luwero, Katikamu, Nangunga, Kimwanyi, Semuto, and Kasawo.

Some cooperatives were established outside the areas best suited to intensive dairying. In these drier, more sparsely populated locations, in which stockkeeping plays a relatively larger role in the agricultural economy, monopolization of the rural trading center's milk market is an insignificant prize, given its small size. To a far greater extent than in the more densely populated, intensive dairying areas, milk marketing operations must from the very beginning be based upon the bulking of milk for transport over long distances to the major urban centers. The possibility of achieving a monopoly position in the local milk market through government backing was less important to cooperatives in such areas than the more concrete forms of assistance which government provided, free or at reduced cost: coolers, transport, and trained staff to receive and dispatch farmers' milk. Once again, of course, such aid was available to cooperatives, but not to private commercial undertakings.

4.   For the sake of brevity, we include crossbred cattle in the term "exotics."

5.   See Sherman, Ugandan Dairy Policy, Chapter 7.

6.   See D.G.R. Belshaw, "Economic Constraints on the Production of Animal Protein in Uganda: Some Policy Implications," in Department of Veterinary Services and Animal Industry, Annual Conference 1969 (Entebbe: Government Printer, 1969), p. 42. Belshaw's evaluation of the situation is somewhat different from that presented here.

7.   Exceptions are a few farmers operating very large estates whose operations are large enough to amortize their own cooling

and transport. The one intensive dairyman in our survey who was not a cooperative member and did not wish to join was an African beneficiary of Amin's "economic war," who had been allocated an Asian estate. He indicated that, as he had his own cooler and trucks, "there is no problem to force me to enter a cooperative."

8. For an interesting literary depiction of the relations between Ganda cultivators and their western herdsmen, see B. Lubega, The Outcasts (London: Heinemann Educational Books, 1971).

9. References to purposive-solidary motives were encountered both in written material and in interviews with society officers. Among the societies for which such references were found were: Bugerere Balunzi, Katikamu, Nangunga, Kimwanyi, Semuto, and Kasawo.

10. An example of first, limited steps toward the use of the cooperative as a basis for general community leadership may be found in the charitable contributions which some of the more successful societies (Katikamu, Kkungu) included in their distribution of surplus.

11. Of the nonofficers in the highest nonfarm bracket, only one-fifth met all three of the following criteria: (a) earned total farm profits equal to or greater than 15 percent of nonfarm income; (b) dwelt all or most of the time on their dairy farm or in the near vicinity; and (c) did not employ a farm manager. By contrast, roughly two-thirds of the society officers met all three of these standards, indicating the greater relative importance of the dairy farm in their economic activities.

12. For this and following references to Mafeje's work, see Audrey I. Richards et al., eds., Subsistence to Commercial Farming in Present Day Buganda (London: Cambridge University Press, 1973), pp. 198-231.

13. The case studies which follow are based upon society files, Veterinary and Cooperative Department files, and interviews with society officers, members, and employees.

14. For the sake of brevity, we use the term "dairy agencies" to refer to the Ministry of Animal Industry/Animal Resources, the Veterinary Department, and the Dairy Corporation. The latter two bodies were responsible to the Ministry and were the administrative units primarily concerned with promotion of milk production and marketing.

15. Societies' treatment of nonmembers' milk was among the major issues considered by committees of inquiry appointed by the Minister of Animal Industry to investigate the operation of a number of milk collection centers. See Chairman/Toro Dairy Cooperative Society to Chairman/Committee of Enquiry, 5 May 1970, and Minister of Animal Industry to Chairman/Toro Dairy Cooperative Society, 22 June 1970, in society files; various letters and memoranda in a special committee on inquiry file, District Cooperative Department Office/Kigezi, "Report of the Ad Hoc Committee

of Enquiry into the Marketing of Milk in Ankole District . . ."
(typescript, December 1968), Co-operative Department/national
headquarters files.

16. Interview, secretary of the Kimwanyi Dairy Cooperative
Society, November 1973.

17. The Co-operative Department did for a time suspend registration of dairy societies. However, this step was taken in response to the dairy agencies' unilateral seizure of a number of
collection centers which had been in cooperative hands; it did
not result from a change in the Department's belief that cooperatives were the proper organizational framework for farmers seeking to market milk.

18. For a fully detailed discussion of the matters presented
below, see Sherman, Ugandan Dairy Policy, Chapter 6.

19. "Speech to be delivered by the Hon. J.K. Babiiha, Vice-
President, Minister of Animal Industry, Game and Fisheries at
opening ceremony of Mityana Milk Cooling Plant on Saturday 26th
November, 1966 at 10:00 AM," in Co-operative Department/national
headquarters files.

20. See Canadian International Development Agency Study Team,
Dairy and Medium-Sized Farm Development in Uganda (Ottawa: CIDA,
1972), pp. 166, 231-34. The figure for cooperative centers would
seem to be somewhat exaggerated. We cannot account for more than
12 centers under cooperative management, so long as the count is
restricted to centers equipped with a government-supplied cooling
device comparable to those installed at Dairy Corporation-oper-
ated sites.

21. Sherman, Ugandan Dairy Policy, pp. 189-203.

22. Ibid., pp. 212-14.

23. "How the Milk is Collected," Uganda Argus, supplement to
the issue of 30 October 1970, p. 3.

24. An abortive attempt to realize the gains of organizing
the dairy agencies' farmer clientele, without incurring the liabilities imposed by cooperatives' independent legal standing and
their ability to rely upon an alternative governmental patron,
was made in 1969. A letter written for the Permanent Secretary
of the Ministry of Animal Industry called upon Regional Veterinary Officers to promote the formation of district Dairy Farmers'
Associations. See Permanent Secretary/Ministry of Animal Industry, Game and Fisheries, to all Regional Veterinary Officers,
4 July 1969, in Veterinary Department/Mukono files.

25. See CIDA Study Team, Dairy Farm Development, Chapter 6.

26. Interview, Chairman-Managing Director/Dairy Board, November 1973; interview, Acting Deputy Commissioner of Veterinary
Services, November 1973.

27. See letters from Chairman-Managing Director/Dairy Corporation to Secretary/Akoraheka Dairy Cooperative Society, 11 May
and 12 May 1972, in Co-operative Department/national headquarters
files.

28. See the letter from General Manager/Dairy Corporation to Kigezi Dairy Cooperative Society, 10 August 1972, in Co-operative Department/national headquarters files.

NOTES TO CHAPTER 8

1. For the concept of organizational ideology, see Anthony Downs, <u>Inside Bureaucracy</u> (Boston: Little, Brown, 1968), Chapter 19.
2. The Registrar of Cooperative Societies, <u>Annual Report for the Year Ended 31st December, 1947</u>; <u>1948</u> (Entebbe: Government Printer); Department of Cooperation, <u>Annual Report . . . 1949</u>; <u>1950</u> (Entebbe: Government Printer). We speak here of the colonial administration; there were some more stirring appeals emanating from Fabian circles influential in the Labour government of 1945-51.
3. See Department of Cooperative Development, "Annual Report . . . 1963-1971," Mimeo. (Kampala: Department of Co-operative Development). Educational materials prepared in the Department speak of the need to eliminate middlemen's profits and the need to prevent exploitation of the small, poor farmer. However, the latter point follows the former without elaboration, and there is nothing in the text to suggest specifically that the conflict referred to might grow out of the opposed interests of weaker and stronger farmers. To the contrary, farmers are told that the society officer must be "successful in his own affairs . . . a good farmer." See Department of Cooperative Development, <u>The Committee</u> (Entebbe: Department of Agriculture, Information and Visual Aids Centre, 1966), p. 16; and Department of Cooperative Development, <u>Uganda Co-operative Education Manual</u> (Kampala: Department of Cooperative Development, 1968?), Lecture no. 5, ¶II/C, ¶II/D. It was not, of course, the bureaucracy which devised the group farm idea.
4. J.B.B. Isabirye, Acting Commissioner for Cooperative Development, "The Duties and Functions of the Department of Co-operative Development," Mimeo. (Kampala: Department of Co-operative Development, September 1973).
5. Ibid., ¶16.
6. Ibid., ¶42.
7. Department of Cooperative Development, "Annual Report . . . 1964," ¶174; "1965," ¶167; "1966," ¶200-¶201; "1969," ¶17.
8. Isabirye, "Duties and Functions," ¶47.
9. Ibid., ¶35, ¶39.
10. See, for example, the annual accounts of the Katikamu and Bugerere Balunzi Dairy Societies in the files of the East Buganda District Cooperative Department headquarters and the national Department headquarters in Kampala.
        Accounting instructions for dairy societies were not

found in the files of the Livestock Section in the Department's national headquarters.

11.   District Cooperative Officer/East Buganda to Cooperative Assistant, through Assistant Cooperative Officer/Bugerere County, 22 March 1973, in East Buganda District Cooperative Department headquarters files.

12.   The decision to allocate coolers only to registered cooperatives was part of the dairy agencies' reversal of attitude toward the involvement of cooperatives in milk collection in the post-coup period.   According to the Acting Chairman/Managing Director of the Dairy Corporation:

> It is declared government policy to encourage the participation of farmers in the marketing of their livestock products, through co-operative societies.   This corporation would do everything possible to assist in achieving this aim, but in doing this, we should abide by the existing cooperative law. And one basic requirement is that, a Co-operative Society must be registered under the co-operative law.

See Acting Chairman/Managing Director of the Dairy Corporation to all Regional Veterinary Officers, 7 January 1972, in the files of the Regional Veterinary Officer/Buganda in Kampala.

13.   Uganda, "The Co-operative Societies Act, 1970," ¶2-¶3.

14.   Assistant Cooperative Officer/Livestock-East Buganda, visit report (1 August 1972); Animal Husbandry Officer/Dairy-Buruli and Bulemezi Counties to Assistant Cooperative Officer/Livestock-East Buganda, 4 August 1972, in East Buganda District Cooperative headquarters files.

15.   The chairman of the Kimwanyi Society was known as a very wealthy businessman, and a Society member served as a committeeman of the East Buganda Co-operative Union.

16.   Isabirye, "Duties and Functions," ¶24a (emphasis added).

17.   See Bye-Laws of the ----- Co-operative Society Ltd., Form no. Co-op. 26.   Paragraph 50 (3) states: "The society shall deal with members only."

18.   Assistant Cooperative Officer/Livestock-East Buganda, report on Kimwanyi special general meeting (30 August 1972), in East Buganda District Cooperative Department headquarters files. The citation is taken from the summary of the District Cooperative Officer's speech.

19.   For an excellent summary of this view, see Bernard B. Schaffer, "The Deadlock in Development Administration," in Colin Leys, ed., Politics and Change in Developing Countries (London: Cambridge University Press, 1969), pp. 177-211.   See also M.J. Esman, "Administrative Doctrine and Developmental Needs," in E.P. Morgan, ed., The Administration of Change in Africa (New York: Dunellen, 1974), pp. 3-26.

20.   Our findings concerning the Uganda Veterinary Department

point in the same direction. For a critical review of the call
for a nonbureaucratic approach to development administration,
growing out of research on the administration of agricultural
extension in Kenya, see David K. Leonard, <u>Reaching the Peasant
Farmer</u> (Chicago: University of Chicago Press, 1977), pp. 217-23.
    21. Nelson Kasfir, "Prismatic Theory and African Administra-
tion," <u>World Politics</u>, 21, no. 2 (January 1969), pp. 307-10.
    22. A review of the East Mengo (later Buganda) District Team
and Planning Commission meeting minutes and Cooperative and Veterinary Department files shows the lack of any substantial intervention in and direction of the activities of the substantive departments by the District Commissioner.
    23. See Leonard, <u>Reaching the Peasant Farmer</u>, pp. 201-09.
    24. T.J. Lowi, <u>The End of Liberalism</u> (New York: W.W. Norton,
1969), Chapter 5.
    25. Contrast Esman, "Doctrine and Needs," and Earl M. Kulp,
<u>Rural Development Planning</u> (New York: Praeger, 1970), Chapters
1-3.
    26. On the issue of professional versus bureaucratic elements
in administration, see Charles Perrow, <u>Complex Organizations</u>
(Glenview, Ill.: Scott, Foresman and Company, 1972), pp. 52-58;
Amitai Etzioni, <u>Modern Organizations</u> (Englewood Cliffs, N.J.:
Prentice-Hall, 1964), Chapter 8.
    27. This is the case for the Cooperative, Veterinary, and
Agriculture Departments. Top officials' careers may be traced
in the staff lists appended to each department's annual reports.
    28. Isabirye, "Duties and Functions," ¶4.

NOTES TO CHAPTER 9

    1. Dennis Austin, <u>Politics in Ghana 1946-1950</u> (London: Ox-
ford University Press, 1964), p. 275.
    2. The definitive study of the Ashanti state is Ivor
Wilkes, <u>Asante in the Nineteenth Century: The Structures and
Evolution of a Political Order</u> (London: Cambridge University
Press, 1975).
    3. Polly Hill, <u>Migrant Cocoa Farmers of Southern Ghana</u> (Lon-
don: Cambridge University Press, 1973), provides the classic ac-
count.
    4. Ibid., pp. 11-86.
    5. F.M. Bourret, <u>Ghana: The Road to Independence</u> (London:
Oxford University Press, 1960), p. 28.
    6. Bjorn Beckman, <u>Organizing the Farmers: Cocoa Politics and
National Development in Ghana</u> (Uppsala: Scandinavian Institute of
African Studies, 1976), pp. 47-48.
    7. Bourret, <u>Ghana</u>, pp. 67-68.
    8. See, for example, David Brokensha, <u>Social Change at Lar-
teh, Ghana</u> (Oxford: Clarendon Press, 1966).
    9. For a thorough treatment of the organization of cocoa

farming, see R.A. Kotey, C. Okali, and B.E. Rourke, eds., The Economics of Cocoa Production and Marketing (Legon: University of Ghana Press, 1974).

10. Austin, Politics in Ghana, p. 5.

11. Bourret, Ghana, p. 150.

12. Ibid., pp. 204-5.

13. Beckman, Organizing the Farmers, pp. 184-96.

14. Bob Fitch and Mary Oppenheimer, Ghana: The End of an Illusion (New York: Monthly Review Press, 1966), pp. 43-44.

15. Kwame Nkrumah, Ghana (Edinburgh: Thomas Nelson and Sons, 1959), p. 179.

16. Austin, Politics in Ghana, p. 60.

17. For a sympathetic political biography of Nkrumah, see Basil Davidson, Black Star: A View of the Life and Times of Kwame Nkrumah (New York: Praeger, 1974).

18. Austin, Politics in Ghana, p. 371.

19. A very useful summary of Nkrumah's economic policy is provided by John Esseks, L'Afrique de l'indépendance politique à l'indépendance économique (Paris: François Maspero, 1975), pp. 91-113.

20. On the external aspects of the socialist phase, see Robert Legvold, Soviet Policy in West Africa (Cambridge, Mass.: Harvard University Press, 1970); and W. Scott Thompson, Ghana's Foreign Policy 1956-1966 (Princeton, N.J.: Princeton University Press, 1969).

21. St. Clair Drake and Leslie Lucy offer a well-done case study of the strike, in Gwendolen Carter, ed., Politics in Africa (New York: Harcourt, Brace and World, 1966), pp. 67-118.

22. For critical appraisals, see Henry Bretton, The Rise and Fall of Kwame Nkrumah (New York: Praeger, 1966); T. Peter Omari, Kwame Nkrumah: The Anatomy of an African Dictatorship (Accra: Ghana Publishing Corporation, 1970).

23. These reports serve as the basis for Victor T. Levine, Political Corruption: The Ghana Case (Stanford, Calif.: Hoover Institution Press, 1975).

24. For coverage of this period, see Dennis Austin and Robin Luckham, eds., Politicians and Soldiers in Ghana 1966-1972 (London: Frank Cass, 1975).

25. Washington Star, 2 September 1978. West Africa, 3191 (11 September 1978), p. 1775; 3183 (17 July 1978), p. 1373.

26. Kotey, Okali, and Rourke, The Economics of Cocoa, p. 53.

NOTES TO CHAPTER 10

1. The following works provided the bulk of information on the pre-World War II cooperatives and quasi-cooperatives in the Gold Coast: G.M. Kay, The Political Economy of Colonialism in Ghana (Cambridge: Cambridge University Press, 1972); S.N. La Any-

one, Ghana Agriculture (London: Oxford University Press, 1963);
J.C. DeGraft-Johnson, African Experiment: Cooperative Agriculture
and Banking in British West Africa (London: Watts, 1958); Margaret Digby, Agriculture Cooperation in the Commonwealth (Oxford:
Blackwell, 1951); H.B. Jeffrey, "Conflict and Cooperation--An
Historical Perspective of Ghanaian Cooperation 1928-1970," Review
of International Cooperation, 72, no. 1 (1980), pp. 260-69.

    2. Polly Hill, Migrant Cocoa Farmers of Southern Ghana (Lon-
don: Cambridge University Press, 1963), and "Ghanaian Capitalist
Cocoa Farmers," in Polly Hill, Studies in Rural Capitalism in
West Africa (Cambridge: University Press, 1970), pp. 21-29.

    3. DeGraft-Johnson, African Experiment.

    4. On the mechanics of cocoa production, see R.A. Kotey, C.
Okali, and B.E. Rourke, The Economics of Cocoa Production and
Marketing (Legon: University of Ghana, 1974); Sara S. Berry, Co-
coa, Custom, and Socio-Economic Change in Rural Western Nigeria
(Oxford: Clarendon Press, 1975).

    5. DeGraft-Johnson, African Experiment.

    6. Ibid.

    7. Kay, Political Economy, pp. 238-40.

    8. Ibid., pp. 253-81, reprinted several sections of the
Nowell Commission, Report of the Commission on the Marketing of
West African Cocoa, 1938.

    9. Details on activities of the Department of Cooperatives
are taken from their Annual Reports from 1944-45 to 1957-58.

    10. Ibid., and La Anyone, Ghana Agriculture, p. 186.

    11. In 1944 the government appointed the Havers Commission
to investigate rural indebtedness. On the basis of their findings, the Commission recommended that credit be made available
at prevailing interest rates as a way to abolish usury in the
countryside. The Andre Jones Memorandum (1945) recommended a
reserve trust established through the Marketing Board to relieve
rural indebtedness. According to the 1957 Report on Rural Indebtedness (Accra: Ministry of Agriculture), three-quarters of
the farmers in southern Ghana and Ashanti are indebted in some
way. The Puckridge Commission (1946) also called for credit to
rural farmers to relieve indebtedness. The pioneer official
study of the cocoa industry, W.H. Beckett, Akokoaso Village (Accra: Department of Agriculture, 1935-1940), found that indebtedness among cocoa producers was already widespread.

    12. Information on the CPP's early efforts to organize cocoa
marketing, and the CPP's relationship with the cooperatives, may
be found in the following sources: Jebowu Commission, Report of
the Commission of Enquiry in the Affairs of the Cocoa Purchasing
Co. Ltd. (Accra: Government Printer, 1956); DeGraft-Johnson, African Experiment. See also the thorough and excellent account
of CPP action in the cocoa field in Bjorn Beckman, Organizing the
Farmers: Cocoa Politics and National Development in Ghana (Uppsala: Scandinavian Institute of African Studies, 1976).

13.   Ibid., pp. 53-54.

14.   Ibid., pp. 65-68.

15.   Report of the Committee of Enquiry on the Local Purchasing of Cocoa (Accra: State Publishing Corporation, 1967), p. 6.

16.   Ghana, Report of the Commission of Enquiry into the Circumstances Leading to the Takeover by the Erstwhile United Ghana Farmers Council Cooperatives of the Business and Assets of the Former Cooperative Societies (Accra: State Publishing Corporation, 1970); Beckman, Organizing the Farmers, pp. 72-73.   The formal recipient of the subsidy was the UGFC-related "National [later Central] Cooperative Council"; a subsidy had earlier been provided to the cooperative apex body, the Alliance of Ghana Cooperatives.

17.   This shift had been long contemplated; see Charles Wilson, The History of Unilever, 2 vols. (London: Cassell, 1954).

18.   Beckman, Organizing the Farmers, pp. 79-84.

19.   The contrast in the roles of the Cooperative Departments in Ghana and Uganda is worthy of note.   The Ghana Department was not able to sustain as paternalistic a tutelary relationship with the cooperatives, in good part because the Ghanaian cooperative leaders were more experienced commercial operators than their Ugandan counterparts.   Nor do we find in Ghana important cooperative managers coming up through the Cooperative Department. While the Ugandan Cooperative Department did not rank high in bureaucratic prestige, it was performing a service valued by government; in Ghana, from the time that Nkrumah became Leader of Government Business in 1951, the Cooperative Department found itself an object of political suspicion.

20.   Beckman, Organizing the Farmers, pp. 91-100.

21.   Kotey, Okali, and Rourke, Economics of Cocoa, pp. 173-74.

NOTES TO CHAPTER 11

1.   General Ankrah was made head of the National Liberation Council, although Colonel Kotoka was the acknowledged leader of the Council.

2.   Ankrah's address to the nation on 24 February 1966, followed by NLC Decree Number 23 on 25 March 1966.

3.   Ghana Cooperative Marketing Association (GCMA) minutes for 9 September 1966; 5 October 1966; 24 January 1967.

4.   Bjorn Beckman, Organizing the Farmers: Cocoa Farmers and National Development in Ghana (Uppsala: Scandinavian Institute of African Studies, 1976), pp. 110-12.

5.   Ibid., pp. 214-18.

6.   Ibid., p. 107.

7.   Interviews with: J.C. DeGraft-Johnson, March 1974; Casely-Hayford, March 1974; M. Oboobi, former registrar, 5 April 1974; F. Mark-Addo, former President of GCMA, 9 March 1974.

8. The complete report was published by the Government of Ghana as <u>Report of the Committee of Enquiry on the Local Purchasing of Cocoa</u> (Accra: State Publishing Corporation, 1967). The report contains a brief chronology of the events leading up to the formation of the Committee of Inquiry.

9. Interviews with DeGraft-Johnson, Casely-Hayford, Mark-Addo, Oboobi. Also, interviews with Boafo, Chief Accountant of GCMA, 23 January 1974, 12 December 1973; Ntim, former Secretary, GCMA, 23 April 1974.

10. GCMA minutes, and previously cited interviews.

11. The NLC was following the recommendation of the DeGraft-Johnson Committee in ordering the CMB to make cash advances available. See DeGraft-Johnson Committee, <u>Report</u>, pp. 35, 62.

12. J.C. DeGraft-Johnson, "Cooperatives in Ghana," Paper presented to the First Conference on Cocoa, Accra, March 1973.

13. This argument is based on the record of the NLC's policies on cooperatives and the cocoa marketing system. The GCMA minutes are filled with unmet requests for government assistance. Interviews with cocoa marketing and cooperative officials support the theory that the NLC adopted a policy of benign neglect.

14. This demand came later, in 1968, and again during the NRC period after the cooperatives came to recognize their limited influence and the need to make their presence felt in policy-making circles. Interviews with Oboobi and Ntim.

15. The cooperatives argued this before the DeGraft-Johnson Committee and in the petitions to the NLC. This has remained the official policy of the movement. In their commitment to competition, they were in sharp contrast to Uganda cooperatives, which wanted a monopoly role.

16. The official title of the Commission was "Commission of Inquiry into the Circumstances Leading to the Takeover by the Erstwhile United Ghana Farmers Council Cooperatives of the Business and Assets of the Former Cooperative Societies." The Committee was led by a prominent barrister, Edward N. Moore.

17. The government delay in responding to the Commission with a White Paper reflected the growing disaffection with the cooperatives, according to popular belief. Interviews with Mark-Addo, Oboobi, Ntim, Boafo. Also, interview with Dodoo, Deputy Registrar, 6 June 1974.

18. The cooperatives made numerous appeals, all of which were rejected. GCMA minutes, 8 October 1966; 8 February 1967; 8 February 1968; 21 March 1968; 4 June 1968; 31 March 1970; 23 September 1969; Letter to the Moore Committee, 13 July 1969; and virtually every committee meeting up to the time of the 1972 suspension of GCMA Management Committee.

19. CMBA, <u>Annual Returns and Final Accounts</u>, 1966-67; 1967-68.

20. Interviews with Boofo, Secretary, Ghana Cooperative Bank, and Chief Accountant, GCMA, December 1973-March 1974.

21. Ibid.
22. Interviews with DeGraft-Johnson, Casely-Hayford.
23. Ibid.
24. This line of argument was made in most of the interviews with former cooperative leaders (Ntim, Mark-Addo, Dumoya).
25. Parliamentary debates, 3 September 1971; 3 August 1971; 12 December 1969; 3 December 1971; 19 March 1970; 18 March 1970; 3 March 1970.
26. The 1970 CMB Board of Directors did not include a cooperative representative. Mark-Addo was removed by the December 1970 reorganization and no other cooperative member was reappointed. In the opinion of the cooperative, the director of the CMB was hostile to the movement.
27. GCMA minutes. Mention of the cooperative financial arrangements was made in virtually every committee meeting between 1968 and 1972.
28. Interviews. Interestingly, most Department of Cooperatives staff officers mentioned Busia's apparent preference for the private LBAs.
29. Interviews with S.T. Moulepe, PCO, 2 March 1973, 14 March 1973; H.T. Partey, PCO, Central Region, 20 March 1973; G.S. Isibal, CO, Central Region, 21 March 1973; Boakye-Amah, SCO, Brong Ahafo, 19 March 1974.
30. This was first noted in the GCMA minutes of 2 January 1967. A more emphatic statement of the problem is found in the minutes of 8 February 1967.
31. DeGraft-Johnson Committee, Report.
32. GCMA, Annual Returns and Final Accounts, 1966-67.
33. Newspaper accounts of the GCMA's indebtedness appeared with some regularity, generally at the beginning and end of each cocoa marketing season. The first press reports on cooperative indebtedness to the CMB appeared in January 1968: The Pioneer, 18 January 1968; 26 January 1968; 1 February 1968; Ghanaian Times, 14 March 1968.
34. GCMA minutes, 8 February 1968, first mentioned the Standard Bank loan; Mark-Addo Annual Report of 31 March 1970.
35. GCMA files (letters and notes) for 19 July 1968 through 30 March 1971.
36. GCMA statement of accounts data, 1968.
37. International Labour Organization, UNDP/78/196, October 1970 (F. Haworth, author); B.J. Youngjohns, "The Cooperative Movement of Ghana," Ministry of Overseas Development (United Kingdom), May 1978; Stanley F. Krause, Agricultural Cooperative Development in Ghana (USAID, 1969).
38. Interview with Don Damoga; many others interviewed made similar points.
39. GCMA minutes for 1971-72. Numerous references were made in the meetings.
40. The ABASCO position on decentralization and secession is

put forward in their petitions, resolutions, and correspondence, from 1968 to 1971.

41. GCMA minutes and correspondence.

42. Interviews with Attrous, Chief Auditor, GCMA, December 1973-January 1974. The controversy about the audit unit is discussed in the following correspondence: 26 October 1966, "The Audit Union," GCMA; 25 November 1966, "Audit of Cooperative Society Accounts," GCMA; 20 January 1967, "Progress Report on the Audit Section," GCMA; GCMA File 146, Audit, 1966-71; 20 May 1967, "Policy on Audit," Department of Cooperatives; 8 May 1968, "Department of Cooperatives Allegations against Audit Staff," GCMA; 1943 (confidential) Memorandum on the Establishment of an Audit Union for the Cooperatives.

43. GCMA minutes on special meeting with Major Kwone Asante, 5 July 1972. Numerous newspaper accounts between 1 January 1972 and 3 October 1972.

NOTES TO CHAPTER 12

1. The Ghana survey was carried out by Tim Rose, who employed a modified version of the survey instrument used by Young and Brett in Uganda. A number of questions were altered somewhat to reflect the differences between cocoa cooperatives, and the cotton-coffee cooperatives in Uganda. Research teams of university students from each survey area administered the pre-tested questionnaire to the farmers in the local language. The proportion of the sample taken from any region was a factor of the proportion of the cooperative membership in the region. As such, Ashanti and Brong Ahafo cooperative unions had two-thirds of the total membership. (The two regions also contribute about 60 percent of Ghana's cocoa production and contain two-thirds of the workforce which lists cocoa farming as its primary occupation. Statistical Yearbook 1970 [Accra: State Publishing Corporation].) The remaining 40 percent of the sample was divided equally between the four remaining cocoa-growing regions: Eastern, Western, Central, and Volta. After determining the sample proportions for each region, we randomly selected one union from each of the four smaller cocoa-growing regions, and three each from Ashanti and Brong Ahafo. The choice of farmer was also a random factor, with 50 to 100 members selected from each union. This required first randomly selecting several societies from each union, and sampling the members at the society level. The turnout and access in some societies was amazing, with up to 100 members easily reached and surveyed. This required the assistance of the village chief and society President. Other societies proved more difficult, requiring Rose and his research assistants to substitute a neighboring society for the one originally designated. In all, 11 unions and 17 societies, and 552 cooperative members

were covered. We surveyed 60 Produce Buying Agency members, but
unlike the cooperatives the PBA selection was far less systematic. Cooperation was far less forthcoming.

2. We used a question designed by Fred Hayward for his study
of political integration in Ghana: "When looking at your overall
economic conditions, do you consider yourself to be about average
with others, above average, average, below average, or much below
average?" Fred M. Hayward, "Perceptions of Well-Being in Ghana,
1970 and 1975," African Studies Review, 22, no. 1 (April 1979),
pp. 109-126.

3. Col. Tekhyi, the military director of the Produce Buying
Agency, deviated from the long-held UGFC/PBA principle of not
awarding member bonuses. The former Secretary-General of the
UGFC and Assistant Director of the PBA, Mr. Appiah-Danquah, actively opposed the bonus policy but was eventually defeated on
the issue.

NOTES TO CHAPTER 13

1. Goren Hyden, Beyond Ujamaa (Berkeley: University of California Press, 1980), pp. 133-34.

2. For a representative illustration of this argument, see
John Saul, State and Revolution in East Africa (New York: Monthly
Review Press, 1979). See also Rural Cooperation in Tanzania
(Dar-es-Salaam: Tanzania Publishing House, 1975).

3. Stephen Quick, Bureaucracy and Rural Socialism (Ph.D.
diss., Stanford University, 1975).

4. Claude Rivière, Guinea: The Mobilization of a People
(Ithaca, N.Y.: Cornell University Press, 1977); Henri de Decker,
Nation et développement communautaire en Guinée et au Sénégal
(The Hague: Mouton, 1967); William I. Jones, Planning and Eco-
nomic Policy: Socialist Mali and Her Neighbors (Washington: Three
Continents Press, 1976).

5. For case studies of ineffectual cooperation, see Edward
J. Schumacher, Politics, Bureaucracy, and Rural Development in
Senegal (Berkeley: University of California Press, 1975); Momodu
S.K. Mamneh, Cooperatives in the Gambia: An Examination of the
Administrative Problems of Cooperative Marketing Unions and Their
Impact on National Economic Development (Ph.D. diss., Rutgers
University, 1975).

6. John Eklund, a cooperative specialist with extensive
third world experience, phrased the matter thus: "In every instance, the cooperative movement is being built from the government outward, with an extension of government in management, direction, planning, and operation. Such procedures appear absolutely essential at the present but do present a critical problem for the future in the development of an independent coopera-

tive movement." Konrad Engelman, ed., Building Cooperative Movements in Developing Countries (New York: Frederick A. Praeger, 1968), pp. 29-30.

7. On Egyptian cooperatives, see James B. Mayfield, Rural Politics in Nasser's Egypt (Austin: University of Texas Press, 1971); San-Eki Nakasaka, "The Agricultural Cooperative in Socialist Egypt," Developing Economies, 3, no. 2 (June 1965), pp. 173-94; Bert Hansen, "Arab Socialism in Egypt," World Development, 3, no. 4 (April 1975), pp. 201-211; Robert Marbro, The Egyptian Economy 1952-1972 (Oxford: Clarendon Press, 1974).

8. The 1969 Tunisian rural crisis, over the effort to impose production cooperatives, is an instructive case in point: Lars Rudebeck, "Developmental Pressure and Political Limits: A Tunisian Example," Journal of Modern African Studies, 8, no. 2 (July 1970), pp. 173-98.

9. Gosta Widstrand, ed., Cooperatives and Rural Development in East Africa (Uppsala: Scandinavian Institute of African Studies, 1970); Goren Hyden, Efficiency versus Distribution in East African Cooperatives (Nairobi: East African Literature Bureau, 1973).

Index

Hansen, Holger Bernt, _Ethnicity and Military Rule in Uganda_ (Scandinavian Institute of African Studies, 1977), 235, 239

Havers Commission, _Report on Rural Indebtedness_ (Ministry of Agriculture, Accra, 1957), 251

Hayden, E.S., "The History of the Bugisu Coffee Scheme" (Mbale, Uganda, 1953), 234, 239

Hayward, Fred M., "A Reassessment of Conventional Wisdom about the Informed Public: National Political Information in Ghana," _American Political Science Review_, 70, 2 (June 1976), 242; "Perceptions of Well-Being in Ghana, 1970 and 1975," _African Studies Review_, 22, 1 (April 1979), 256

Helleiner, G.K., "Economic Change and Rehabilitation in Uganda," annual meetings, African Studies Association, 1980, 236

Hickey, G.C., _Village in Vietnam_ (Yale University Press, 1964), 231

Hill, Polly, _Migrant Cocoa Farmers of Southern Ghana_ (Cambridge University Press, 1973), 249, 251

Hindley, Donald, _The Communist Party of Indonesia 1951-1963_ (University of California Press, 1964), 232

Hoima, 70

Hunt, Diana: 88, 89; _Credit for Agricultural Development: A Case Study of Uganda_ (East African Publishing House, 1975), 231, 240, 241; "The Ugandan Cooperative Credit Scheme," _East African Journal of Rural Development_, 5, 1-2 (1972), 231, 241

Hunter, Guy, _The Administration of Agricultural Development_ (Oxford University Press, 1970), 233; "Agricultural Administration and Institutions," Conference on Strategies of Agricultural Development in the 1970s, Food Research Institute, Stanford University, 1971, 227; _Modernizing Peasant Societies_ (Oxford University Press, 1969), 230; and A.H. Bunting and Anthony Botral, _Policy and Practice in Rural Development_ (Allanheld, Osmun & Co., 1976), 251

Huntington, Samuel: 15; _Political Order in Changing Societies_ (Yale University Press, 1968), 229

Hyden, Goran, _Beyond Ujamaa_ (University of California Press, 1980), 227, 232, 256; _Efficiency Versus Distribution in East African Cooperatives_ (East African Literature Bureau, 1973), 228-29, 257

Ideology, cooperative: 4, 6-9, 25, 222; in Cooperative Department, 151-52

Imperial British East African Company, 33

Incentives: for cooperation, 19-21, 23, 26, 102, 105-7, 178; for dairy cooperatives, 122-27, 136-37, 140, 142

India, 9

Indigenous cooperation, 14

Innovation: 21, 22, 39, 94; in dairy farming, 135-36

Martin, David, _General Amin_
(Faber and Faber, 1974),
235, 236
Masaka, 80
Masefield, G.B., _Agricultural_
Change in Uganda 1945-1960
(Food Research Institute,
Stanford University, 1962),
234
Masindi, 70
Mayanja, Abu, 73
Mayfield, James B., _Rural_
Politics in Nasser's Egypt
(University of Texas Press,
1971), 257
Mazrui, Ali, _Soldiers and_
Kinsmen in Uganda (Sage
Publications, 1975), 233,
235
Mbale, 70, 77, 115
Mechanization of farming, 98
Media, exposure to, 104
Membership, cooperative, 13,
66, 105, 127, 158, 208
Merhav, Meir, _Technological_
Dependence, Monopoly, and
Growth (Pergamon Press,
1969), 230
Michels, Roberto, 8
Middlemen, in trade, 18-19,
41, 151, 176
Migdal, Joel, _Peasants, Poli-_
tics and Revolution (Prince-
ton University Press, 1974),
230; "Policy and Power: A
Framework for the Study of
Cooperative Policy Contexts
in Third World Countries,"
Public Policy, 25, 2 (Spring
1977), 229
Miles, J.G., 2
Milk collection, 144-45
Ministry of Animal Resources
(Uganda), _Farmers' Forum Re-_
port 1971 (Government Print-
er, Entebbe, 1973), 241
Ministry of Information and
Broadcasting (Uganda), _The_
First 366 Days (Government

Printer, Entebbe, 1972?),
240
Mismanagement, 109-10
Missionaries, 33, 34, 75-76,
163
Mittelman, James H., _Ideology_
and Politics in Uganda
(Cornell University Press,
1975), 233
Moore, Edward, _Commission of_
Inquiry into the Circum-
stances Leading to the Take-
over by the Erstwhile United
Ghana Farmers Council Coop-
eratives of the Business and
Assets of the Former Cooper-
ative Societies (Government
Printer, 1968), 192, 195,
253
Moris, John R., _The Agrarian_
Revolution in Central Kenya:
A Study of Farm Innovation
in Embu District (Ph.D.
diss., Northwestern Univer-
sity, 1970), 233
Mortimer, Rex, _Indonesian Com-_
munism under Sukarno: Ideol-
ogy and Politics 1959-1965
(Cornell University Press,
1974), 232
Mount Elgon, 38, 74
Muduku, S.G., 78, 79
Mukwaya, A.B., "The Rise of
the Uganda African Farmers'
Union in Buganda" (East Af-
rican Institute of Social
Research, n.d.), 238
Musazi, Ignatius, 48, 60, 71-
72, 73, 217
Muslims, Uganda, 33
Mutenyo, S.K., 83
Mutesa II, Kabaka of Buganda.
See Kabaka of Buganda

Nairobi, 77
Nakasaka, San-Eki, "The Agricultural Cooperative in Socialist Egypt," _Developing_

Library of Congress Cataloging in Publication Data

Young, Crawford, 1931-
   Cooperatives & development.

   Includes index.
   1. Agriculture, Cooperative--Ghana.   2. Agriculture,
Cooperative--Uganda.   3. Agriculture and state--Ghana.
4. Agriculture and state--Uganda.   I. Sherman, Neal P.
II. Rose, Tim H.   III. Title.   IV. Title: Cooperatives
and development.
HD1491.G4Y68   334'.683'09667   81-50830
ISBN 0-299-08710-7   AACR2